W9-BXW-684

Tilapia buttikoferi

A FISHKEEPER'S GUIDE TO
AFRICAN CICHLIDS

Lamprologus pulcher

Julidochromis ornatus

A FISHKEEPER'S GUIDE TO

AFRICAN CICHLIDS

A splendid introduction to this diverse and attractive
group of tropical freshwater fishes

Dr Paul V Loiselle

Tetra Press

No. 16037

A Salamander Book

© 1988 Salamander Books Ltd.,
Published by Tetra Press,
201 Tabor Road,
Morris Plains, NJ 07950.

ISBN 3-923880-39-1

This book may not be sold outside the
United States of America and Canada.

All rights reserved. No part of this book
may be reproduced, stored in a retrieval system or transmitted
in any form or by any means, electronic, mechanical,
photocopying, recording or otherwise, without the
prior permission of Salamander Books Ltd.

All correspondence concerning the content of this volume
should be addressed to Tetra Press.

Astatotilapia sp.

Credits

Editor: Vera Rogers   Design: Graeme Campbell
Colour reproductions:
Bantam Litho Ltd.
Filmset: SX Composing Ltd.
Printed in Belgium by Henri Proost & Cie, Turnhout.

Author

Dr Paul Loiselle's interest in aquarium fishes dates back over twenty years. He took his Master's Degree at Occidental College in Los Angeles and obtained his doctorate from the University of California at Berkeley. During five years as a Peace Corps fisheries biologist in West Africa, he carried out faunal and environmental surveys in Togo and Ghana and has since returned to Africa to make first-hand observations of cichlid behaviour in Lakes Victoria and Tanganyika. He is a founder member and Fellow of the American Cichlid Association and Technical Editor of the A.C.A.'s Journal. Outside the USA he has reached a wide audience of aquarists as the author of books and numerous articles, published internationally, on the care and breeding of aquarium fishes.

Contents

Introduction

African cichlids are some of the most colourful and exciting freshwater aquarium fishes and each year new species are introduced to fishkeeping. This happy state of affairs arises from the simple fact that there are so many of them. Lake Malawi alone boasts more cichlid species than all of South America, while the total number of species described to date from Lakes Tanganyika and Victoria exceeds that native to the entire New World! Aquarists are thus assured of finding a cichlid appropriate to their circumstances, but the task of surveying the group with precision is correspondingly magnified. By the most conservative estimate, at least six distinct lineages are represented in the ranks of those cichlids of interest to aquarists. These fish vary markedly in terms of their maintenance requirements and have strikingly different mating systems and methods of reproduction. This is part of their fascination, but it can create confusion.

To make any sense at all of the cichlids of Africa, we must divide them into smaller, more manageable groups. This exercise can be based on many different criteria, but from a practical point of view, the most helpful way of classifying African cichlids is to ask 'where do they originate?' and 'what mating system do they employ?' The answer to the first question dictates the water conditions required by a given species in order to prosper in captivity and provides some insight into how its quarters should be furnished. The second answer influences the amount of space required by these fishes in the aquarium and dictates how best to manage spawning efforts. In Part One of this book we endeavour to answer these questions by examining all aspects of keeping these fishes, with a particular emphasis on their water requirements, tank selection and aquascaping, feeding and routine maintenance, breeding and rearing.

General considerations

As suggested in the Introduction (page 10-11), it is worth considering in some detail the natural habitats and spawning behaviour of the major subgroups of African cichlids, in order to achieve a better understanding of their maintenance requirements in captivity. For practical purposes, it is convenient to recognize three geographical subsets of African cichlids. The first group consists of those species native to the rivers of Africa. Cichlids of the northern Great Lakes, such as Lake Victoria can be included in this group because they require similar general care. A quick glance at the accompanying map will reveal that riverine cichlids are found over a tremendous area, which is divided into a number of discrete regions by ichthyologists. The second major group consists of cichlids native to Lake Malawi and in the third group are those cichlids found only in Lake Tanganyika.

Riverine cichlids
Here it is important to distinguish between rivers that flow for most, or all, of their length through savanna (grassland) habitats, and those that drain forested regions.

Savannas are characterized by highly seasonal patterns of rainfall. The rainy season is brief; a year's worth of rain falls over the course of three or four months. As a result, savanna rivers vary enormously in their rate of flow and chemical make-up during the course of one year. In the rainy season they become veritable torrents that may leave their beds and inundate enormous tracts of land. At this time, pH values in the main channel of the river hover closely around 7.0 (i.e. neutral). Depending upon local conditions, they can drop as low as 6.0 (i.e. acidic) in quiet-water habitats on the river's flood plain. Due to the tremendous influx of rainwater, hardness values seldom exceed 3°dH (see panel page 14).

During the dry season, however, conditions are dramatically

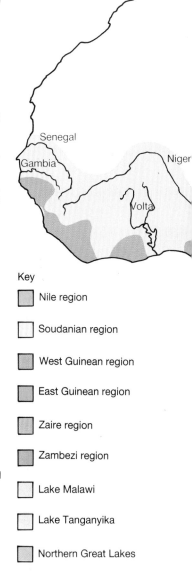

Key

Nile region

Soudanian region

West Guinean region

East Guinean region

Zaire region

Zambezi region

Lake Malawi

Lake Tanganyika

Northern Great Lakes

Above: *The coloured regions indicated on the map and explained in the key are home to most of the aquarium fishes native to the continent, as well as to all of the cichlids covered in this book.*

Distribution of African fish species by region

Basic water chemistry: pH and hardness

The pH of water
The degree of acidity or alkalinity of water is expressed in terms of pH value, which literally means 'hydrogen power'. The scale is based inversely on the concentration of hydrogen ions in the water; the more hydrogen ions, the more acid the water and lower the pH value. The pH scale ranges from 0 (extremely acidic) to 14 (extremely alkaline), with a pH value of 7 as the neutral point.

pH scale

0	1	2	3	4	5	6	7	8	9	10	11	12	13	14

extremely acidic   neutral   extremely alkaline

The scale is logarithmic, which means that a pH value of 8 represents a ten-fold decrease in the hydrogen ion concentration compared to a pH value of 7. An apparently small change in pH, from say 6 to 8, therefore represents a hundred-fold decrease in the hydrogen ion concentration, which can cause severe stress to many fishes. A number of different tests for measuring pH are available. These include paper strip indicators, liquid pH test kits and more sophisticated electronic pH meters.

Water hardness
Water hardness is related to the amounts of dissolved salts present in the water. Two types of hardness are important to the fishkeeper; *total* or *general hardness* (or GH), which is related to the levels of calcium, and magnesium in the water, and *carbonate hardness* (or KH), which is related to the amounts of carbonate/bicarbonate present. Water hardness is measured by several different scales, including degrees of German hardness (dH°). On this scale, water with a hardness value of 3°dH or less is termed 'soft' (i.e. low in dissolved salts) and water with a hardness value of over 25°dH is termed 'very hard' (i.e. rich in dissolved salts). An alternative scale is based on milligrams of calcium carbonate ($CaCO_3$) per litre, also expressed as parts per million (ppm). Again, test kits are available from aquarium dealers.

Water hardness in comparative terms

dH°	Mg/litre $CaCO_3$	Considered as:
3	0-50	Soft
3-6	50-100	Moderately soft
6-12	100-200	Slightly hard
12-18	200-300	Moderately hard
18-25	300-450	Hard
Over 25	Over 450	Very hard

different. Even large rivers, such as the Niger and the Zambezi, may cease to flow altogether along their upper and middle reaches. What was formerly a mighty torrent is reduced to a series of pools connected by a trickle of water! Evaporation concentrates dissolved substances significantly, so total hardness values as high as

10°dH are not unusual. As a rule, pH values fall somewhere between 7.0 and 7.8 in unplanted habitats but can soar as high as 9.0 (i.e. alkaline) during the day, in heavily vegetated oxbows (the water within a bend in the river), or in a peripheral swamp.

Obviously, fishes that can prosper in such a setting must possess considerable resilience in the face of fluctuating pH and hardness values. Nor are they likely to be very demanding with regard to the chemical make-up of their water in captivity, as long as extremes of pH or hardness are avoided. Not surprisingly, cichlids native to these habitats, such as *Tilapia zillii*, *Oreochromis mossambicus*, *Chromidotilapia guntheri* or *Haplochromis callipterus* have a well-established reputation for hardiness among enthusiastic aquarists!

In forested regions the situation is quite different. A tropical rain forest occurs only where there is more or less constant rainfall. The 'dry' season in such areas simply refers to the time of the year when less rain falls, contrary to the case on the savanna. Forest rivers thus enjoy a much steadier flow and are not characterized by dramatic seasonal variation in pH or hardness values.

An equally important point to remember is that forest streams typically flow over very old rocks, from which any soluble materials have long since been leached. Their waters are thus characterized by very low total and carbonate hardness; values in excess of 2°dH would be considered remarkable under most circumstances. When very soft waters come into contact with significant deposits of decomposing plant matter, they pick up large quantities of organic acids. Such 'blackwater' streams – so-called because of the characteristic coloration of their waters – can have pH values as low as 4.0. Even in so-called 'clearwater' streams, pH values between 5.5 and 6.2 are commonly encountered, while those in excess of 7.0 are virtually unheard of.

Interestingly enough, most cichlids native to forest streams do not seem to object to moderately hard, neutral to slightly alkaline water conditions in captivity. Indeed, a number of blackwater species, such as *Anomalochromis thomasi* and *Hemichromis cristatus*, grow larger – and spawn bigger clutches – under such conditions than they do when an effort is made to duplicate their natural surroundings. What none of these forest-associated riverine cichlids tolerate well are *abrupt* changes in pH or hardness.

Lake cichlids
Under this heading we must differentiate between conditions in the northern Great Lakes and those that prevail in Lake Malawi and Lake Tanganyika.

In Lake Victoria and other northern Great Lakes, water conditions resemble those characteristic of large, savanna-draining rivers during the rainy season. Lake Kivu, however, is characterized by hard alkaline water conditions, comparable to those of the two best known of the African Rift Lakes, Tanganyika and Malawi. Nevertheless, cichlids from Lakes Victoria and Kivu will do well in neutral to slightly alkaline water over a considerable range of hardness values. Bear in mind that, once again, you should avoid abrupt changes in the chemical composition of their water. In this respect, they resemble their distant Tanganyikan and Malawian kin, rather than the cichlids found in rivers with a seasonally variable regime.

Lakes Tanganyika and Malawi are distinctive in that their catchment areas have been geologically active in the relatively recent past. Their rocks are thus much younger than those found in most of western Africa and correspondingly richer in soluble minerals. Additionally, for most of their geological histories, both lake basins have been closed, i.e. with no outlets to the sea. Given the combination of mineral-laden streams flowing into the lakes and the opportunity for concentration of dissolved substances through evaporation over many millions of year, it is no wonder that the waters of both lakes are alkaline and very hard.

Recorded pH values range from 7.7 to 8.6 in Lake Malawi, and from 7.3 to 8.0 in Lake Tanganyika. Values of 10-12°dH total hardness

Below: *In solution, compounds split up (ionize) into negatively charged ions (anions) and positively charged ions (cations). In water, magnesium chloride will form Mg^{++} and Cl^- ions. The pie chart illustrates the relative abundance of the main cations and anions in Lakes Tanganyika and Malawi.*

are recorded for Lake Tanganyika. The waters of Lake Malawi are somewhat less mineralized; here, overall hardness values range from 6-10°dH.

Of equal significance to the aquarist is the fact that both lakes are remarkably stable in terms of chemical composition and water temperature. Neither attribute should come as a great surprise; a volume of water sufficient to fill a basin several hundred kilometres long and over a kilometre (0.62 miles) deep constitutes a formidable heat sink that would require a deluge of Noahchian proportions to influence its pH or hardness. However well adapted their ancestors might have been to the vicissitudes of living in a seasonally variable habitat, cichlids native to these inland seas cope poorly with abrupt environmental changes. Tanganyikan cichlids are particularly sensitive in this respect. Over the course of that

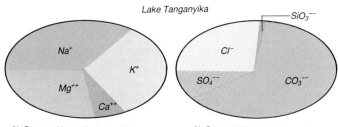

Lake Tanganyika

% Composition principle cations

% Composition principle anions

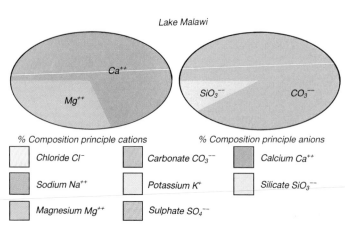

Lake Malawi

% Composition principle cations

% Composition principle anions

☐ Chloride Cl^-

☐ Sodium Na^{++}

☐ Magnesium Mg^{++}

☐ Carbonate CO_3^{--}

☐ Potassium K^+

☐ Sulphate SO_4^{--}

☐ Calcium Ca^{++}

☐ Silicate SiO_3^{--}

16

Above: The Napoleon Gulf of Lake Victoria. Cichlids from this region appreciate neutral to slightly alkaline water and tolerate a range of hardness values.

lake's longer geological history, its indigenous cichlids have had more time to lose the physiological mechanisms that allow their riverine counterparts to cope with seasonal or even diurnal changes in temperature, pH and hardness.

It is worth noting at this juncture that neither lake is noticeably brackish. While an argument can be made on other grounds for adding table salt to the Malawian or Tanganyikan cichlid tank, it cannot be justified on the grounds of recreating a natural habitat. Experience has also shown that Malawian cichlids will prosper in water considerably harder and more alkaline than that of their native lake, while Tanganyikan cichlids will live and breed under softer, less alkaline conditions. Their only absolute requirements are that the pH in their tanks be alkaline and that they be gradually acclimatized to the conditions under which they are to live.

Mating systems
The second major question posed in the Introduction concerned the mating systems employed by different cichlids. The term 'mating system' refers to the nature of the association between the sexes during reproduction. Failure to understand the practical implications of the various types of mating systems – in terms of the housing and breeding requirements of the different groups of fishes – is the single most important reason for lack of success in maintaining and breeding cichlids in captivity. Cichlids are either monogamous or polygamous and here we examine the salient characteristics of each group. The subject of breeding and rearing African cichlids in captivity is discussed in greater detail on pages 54 to 63.

Monogamous cichlids
Among animals that practise monogamy, a single male and a single female collaborate for the duration of a single reproductive effort. Among cichlids, monogamy typically entails sharing the defence of a breeding territory and joint, long-term protection of the mobile fry against predators. Such cichlids are characterized by a pair bond, based upon the ability of each member of the pair to recognize the other, which allows two aggressive animals to cohabit peacefully during the breeding and rearing period.

Above: *The floodplain of the lower reaches of the Zio River in Togo during the rainy season. In the dry season the waters evaporate, hardness values increase and pH levels fluctuate.*

Polygamous cichlids

Among animals that practise polygamy, one individual may consort with several sexual partners during the course of a single reproductive effort. The two most common variations of polygamy encountered among cichlids are harem polygyny and open polygamy. In the first instance, a single male will monopolize access to the same group of females, spawning with each in turn on a recurrent basis. Thus, most harem polygynists possess, at least in rudimentary form, mechanisms that facilitate long-term association of the sexes. However, the degree of male parental involvement depends upon environmental circumstances and on the behavioural idiosyncracies of the particular species in question.

Among openly polygamous cichlids, however, the association of the sexes is limited to the spawning act itself. In nature, it is common for a single individual, regardless of sex, to consort with several partners over the course of a single reproductive effort. Behavioural ecologists recognize different forms of open polygamy, but the feature that is common to all of them – and that is of most significance to the fishkeeper – is that none of the cichlids that practice open polygamy possess behavioural mechanisms that permit the two sexes to share a territory over an extended interval without incident.

These different mating systems clearly have a bearing on how prospective breeders should handle the various cichlids. However, it is also important to take into account a given species' mating system when allocating it living space. For example, reproductively active, monogamous, substrate-spawning cichlids exclude other fishes from a sizeable territory for up to two months at a time. A tank intended to house a community of such cichlids must be sufficiently large to meet the needs of at least one active breeding pair, as well as the requirements of the other aquarium inhabitants seeking a refuge beyond the limits of the breeding territory. Males of polygamous mouthbrooding cichlids, on the other hand, defend smaller territories for a relatively brief interval before spawning. Given a tank of specific volume, the choice is between keeping a greater number of polygamous mouthbrooding cichlids, or a smaller number of monogamous

Mating system and spawning behaviour of African cichlids

Species	See key	Spawning behaviour
Rivers and Northern Great Lakes		
Anomalochromis	M	Semi-open spawner
Astatoreochromis	OP	Mouthbrooder
Chromidotilapia	M	Mouthbrooder
Haplochromis and allied genera	OP	Open spawner
Hemichromis	M	Open spawner
Nanochromis	M/HP	Cave spawner
Oreochromis	OP	Mouthbrooder
Pelvicachromis	M	Cave spawner
Pseudocrenilabrus	OP	Mouthbrooder
Sarotherodon	M	Mouthbrooder
Steatocranus	M	Cave spawner
Teleogramma	HP	Cave spawner
Thysia	M	Semi-open spawner
Tilapia	M	Semi-open spawner

Species	Key	Spawning behaviour	Species	Key	Spawning behaviour
Lake Tanganyika			**Lake Tanganyika cont.**		
Callochromis	OP	Mouthbrooder	Telmatochromis	M/HP	Cave spawner
Chalinochromis	M	Cave spawner	Tropheus moorii	OP	Mouthbrooder
Cyathopharynx	OP	Mouthbrooder	Xenotilapia	OP	Mouthbrooder
Cyphotilapia	OP	Mouthbrooder			
Cyprichromis	OP	Mouthbrooder	**Lake Malawi**		
Eretmodus	M	Mouthbrooder	Aulonacara	OP	Mouthbrooder
Julidochromis	M	Cave spawner	Chilotilapia	OP	Mouthbrooder
Lamprologus	M/HP	Cave spawner	Haplochromis	OP	Mouthbrooder
Ophthalmotilapia	OP	Mouthbrooder	Iodotropheus	OP	Mouthbrooder
Petrochromis	OP	Cave spawner	Labeotropheus	OP	Mouthbrooder
Simochromis	OP	Cave spawner	Labidochromis	OP	Mouthbrooder
Spathodus	M	Mouthbrooder	Melanochromis	OP	Mouthbrooder
Tanganicodus	M	Mouthbrooder	Pseudotropheus	OP	Mouthbrooder

Key – *Mating system:* M= Monogamy; OP=Open polygamy; HP=Harem polygyny

cichlids that share responsibility for raising their fry.

It is interesting to compare the great variety of water requirements and mating systems found among the African cichlids with the demands of Central American cichlids. By way of contrast, the cichlids of Central America, with two exceptions, represent a single evolutionary lineage, prefer the

same water chemistry, thrive over the same temperature range and share the same mating system and method of reproduction.

Below: *The Tanganyikan cave spawner* Telmatochromis bifrenatus, *is monogamous in nature. In captivity, some naturally monogamous species may shift to a polygynous system.*

Water requirements and filtration

With the exception of specialized species that live among the rapids, such as *Steatocranus casuarius, Teleogramma brichardi* and *Haplochromis bakongo*, African cichlids do not relish strong water movement. Most riverine species seek out areas of moderate water flow and, apart from species that inhabit the surge zone, lake cichlids are also found in areas of relative calm. In captivity, African cichlids find the degree of water movement provided by a working filter satisfactory. Even current-loving cichlids adapt to these conditions, though one may suspect that the antisocial behaviour of which many stand accused represents an outlet for energy normally spent fighting the current of their native rivers!

Other than in very densely populated aquariums, the return flow of an outside power or canister filter creates sufficient surface turbulence to facilitate the free exchange of gases from solution. However, filters have been known to suffer temporary failure if they become blocked, or if the siphon action is interrupted. The respiratory needs of African cichlids are proportional to their size, and cichlid communities are often densely populated. The consequences of filter failure can be swift and irreversible, so it is worthwhile including supplementary aeration in a cichlid tank. A small diaphragm pump and an airstone will suffice to insure against possible disaster.

Creating appropriate water conditions

You will find it easier to provide your cichlids with suitable water conditions if you elect to work with species that live comfortably in your tapwater. In the interests of making life a bit easier for aquarists whose preference in cichlids does not march in step with the properties of their local water supply, a brief discussion on the subject of water chemistry seems appropriate.

Hardness and pH value

Many aquarists invest substantial effort in trying to modify the pH of their tapwater. Unless this endeavour is preceded by a successful attempt to alter its hardness, it is at best an exercise in futility and, at worst, is likely to prove a positive hazard to their fish. With rare exceptions, the dominant dissolved chemical substance in fresh water is calcium (more rarely, magnesium) carbonate. The relative abundance

Below: *A liquid pH test kit. Add a few drops of the indicator to a sample of water and compare the colour change to a coloured chart.*

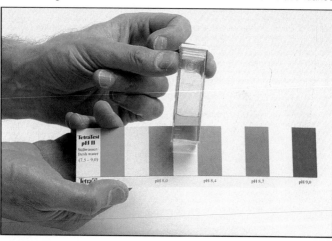

of carbonate ion largely determines the pH of water. Furthermore, its natural buffering capacity sets precise limits on the extent to which the pH can be permanently altered. Thus, aquarists find it useful to ascertain not only the total or general hardness/GH (see page 14), but also the carbonate hardness/KH – a measure of the amount of dissolved carbonate present in the water supply. The ability to measure accurately both pH and hardness is essential to any attempt to modify the chemical composition of aquarium water; reliable test kits are essential to the success of such efforts.

Making water soft

Water with an elevated carbonate hardness is extremely resistant to acidification. It is possible to push its pH into the acid range using sodium biphosphate, but within a few hours it will return to its original value. The pH level of such carbonate-rich waters can be permanently lowered only by 'breaking' the natural buffering system by the gradual addition of a strong acid, such as phosphoric or hydrochloric acid. Alternatively, you can remove dissolved carbonates by passing the water through a selective ion-exchange resin, or a reverse osmosis treatment unit.

Once this has been done, it is a simple matter to lower the pH by progressive addition of commercially available tannic and humic acid concentrates, or by filtering it through acidic peat. Alternatively, you can bypass the problem of hard tapwater completely by using commercially available demineralized water when setting up your tank. Adding 1gm of aquarium salt per 40 litres (approximately 1 level teaspoon per 10 gallons) will raise its concentration of 'physiological ions' to a level that will support fish life without adding to its carbonate hardness. It can then be acidified using either of the techniques previously suggested.

Demineralized water is not cheap, but the cost of it may be acceptable on a one-time basis when setting up an aquarium. However, few aquarists will be willing to buy it in sufficient quantities to carry out regular partial water changes. Techniques exist to reduce the frequency of such changes, but there is no way to eliminate them entirely from a workable nitrogen cycle management programme. It is clear, therefore, that if you wish to keep softwater cichlids in an area with hard tapwater, you would do well to invest in a water treatment system that will supply water of appropriate chemical composition in quantity and at reasonable cost.

In many parts of the world, homes are equipped with commercial water softening systems that use selective ion-exchange resins. These materials substitute sodium (Na^+) for calcium (Ca^{++}) and magnesium (Mg^{++}) ions, and chloride (Cl^-) for carbonate (CO_3^{--}) and sulphate (SO_4^{--}) ions. (The positively charged ions are called cations and the negatively charged ions are known as anions). Water treated in this way is suitable for most domestic uses, but less acceptable for fishkeeping purposes. Ion-exchange resins exist that will completely demineralize tapwater. Such resins substitute hydrogen ions (H^+) for dissolved cations and hydroxyl ions (OH^-) for dissolved anions. These resins are not generally employed in home water treatment systems, since they must be regenerated with strong acids and bases, a hazardous and uneconomical procedure.

The most attractive alternative available to aquarists for softening hard water is a small reverse osmosis water treatment unit. Such systems remove minerals from solution by passing tapwater under pressure through membranes selectively impermeable to dissolved substances. Depending upon the level of use – and on the concentration of the dissolved substances they are called upon to remove – such membranes enjoy a working life of three months to a year, and replacing them is a simple matter.

Reverse osmosis units are more compact and efficient than those relying upon selective ion-exchange resins. They are also much safer, as they do not require treatment with corrosive acids and bases to regenerate their active components. Commercial units capable of producing about 100 litres(22 Imp. gallons/25 US gallons) of demineralized water daily are available at considerably less cost than a water treatment system based on selective ion-exchange resins. The initial outlay may still seem high, but given the cost of bottled demineralized water, such a unit will pay back its purchase price quite rapidly.

Making water hard
Aquarists who wish to keep hardwater cichlids in areas with soft tapwater have an easier time. In this case the process entails adding, rather than removing, salts from solution and it is a much simpler matter to make soft water hard than the reverse. This procedure also automatically raises the pH value of treated water. In areas with moderately soft tapwater, it is enough to filter the tank water through a bed of calcareous material to bring pH and hardness up to acceptable levels. Coral gravel or crushed oyster shell serve this purpose admirably, but dolomite gravel is not recommended. It is only slightly soluble and has the wrong chemical composition to reproduce satisfactorily the water conditions prevailing in either Lake Tanganyika or Lake Malawi.

Remember that it takes time to dissolve sufficient mineral salts from such media to raise the pH value of freshly added water to acceptable levels. To avoid abrupt drops in pH or hardness levels in tanks that rely exclusively on this means of maintaining alkaline

water conditions, do not carry out water changes of more than 25% of the total water volume at any one time.

Confronted with very soft tapwater, you may find it worth the expense of using commercial salt mixes to create suitable living conditions for Rift Lake cichlids. Used as recommended by the manufacturer, these products will consistently yield satisfactory results. Because they contain a high proportion of carbonates, these mixes take quite some time to dissolve completely – up to three days is not unusual. Take this fact into account, both when setting up an aquarium for the first time and when planning water changes. Remember also, that at every water change you must add a quantity of salt mix to the system, equivalent to the amount lost from the tank, in order to maintain pH and hardness at the desired level.

Ammonia, nitrites and nitrates
Fishes obtain the energy necessary to sustain their metabolisms by ingesting and breaking down organic compounds. The breakdown of fats and carbohydrates produces waste products in the form of water and carbon dioxide. Protein digestion yields both of these compounds and, ultimately, nitrogen gas. However, the digestion of protein foods by a fish is only the first step in a long and complex process that is completed in the surrounding environment.

Two of the intermediate products of this process, ammonia and nitrite ion, are known to be toxic to fishes and other aquatic organisms. Ammonia exists in two distinct forms: highly toxic, electrically neutral dissolved ammonia (NH_3), and the harmless, positively charged ammonium ion (NH_4^+). The percentage of each chemical substance depends upon the pH value of the aquarium water. The amount of toxic ammonia increases as tank conditions become more alkaline.

While Malawian and Tanganyikan cichlid keepers must worry about losing fish from ammonia poisoning, such an eventuality is much less likely to befall aquarists who work with riverine species. However, nitrite is almost as dangerous under freshwater conditions and, unlike ammonia, the risk it poses to aquarium fishes is not a function of pH. Long-term exposure to high concentrations of a third nitrogen cycle intermediary – nitrate (NO_3^-) – is suspected of having a harmful effect on both growth and the readiness of a wide selection of ornamental fishes to breed in captivity. The secret of successful cichlid husbandry is to keep concentrations of all three dissolved metabolites (literally substances produced by metabolism) as low as possible.

Regardless of where they originate, African cichlids do not relish long-term exposure to dissolved metabolic wastes.

Species native to the Great Lakes and those inhabiting forest streams or the permanent rapids of large rivers cannot tolerate even brief exposure to either ammonia or nitrite. Neither substance ever builds up to physiologically stressful concentrations in their native habitats and so these fishes have lost any ability their ancestors might have possessed to cope with such stress. Even less than lethal long-term exposure to levels of these substances erodes their natural ability to resist disease and typically leads to death from systemic bacterial infections such as 'Malawi bloat'. Cichlids from seasonally variable riverine habitats are often exposed to higher nitrite and nitrate concentrations for brief intervals during the dry season. They are thus better able to tolerate short-term exposure to these pollutants in captivity, but even they cannot prosper without due attention to nitrogen cycle management.

The nitrogen cycle

The natural processes of the nitrogen cycle convert the dangerous ammonia formed by the decompostion of fish wastes and uneaten food into less toxic products, such as nitrates, which are used as a plant food. Nitrifying bacteria (such as Nitrosomonas *sp.) combine the ammonia with oxygen in the water to form slightly less toxic nitrites. Further bacteria (such as* Nitrobacter *sp.) continue the cycle; with the addition of further oxygen, nitrites are converted into even less harmful nitrates. A plentiful supply of oxygen in the water is essential if the helpful bacteria are to survive and multiply. Provide adequate aeration and never turn off filters for long periods.*

Fish wastes

Food

Uneaten food

Decomposers (Fungi and bacteria)

Plant fragments

AEROBIC CONDITIONS

Ammonia (NH_3/NH_4^+)

Nitrates (NO_3^-)

Nitrite bacteria

Nitrites (NO_2^-)

Nitrate bacteria

Denitrification by anaerobic bacteria

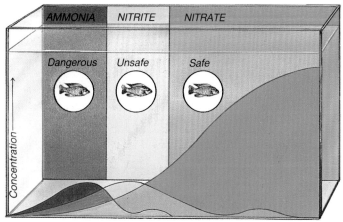

AMMONIA	NITRITE	NITRATE
Dangerous	Unsafe	Safe

Time ⟶   Approximately seven weeks

Above: *In a newly established tank, toxic ammonia and nitrite are produced in overlapping peaks and can reach quite high levels. It may be several weeks before* *nitrifying bacteria responsible for the breakdown of harmful metabolites are present in sufficient numbers. Introduce new fishes gradually.*

Biological filtration
The nitrifying bacteria that convert ammonia to nitrite and thence to nitrate are present in any aquatic habitat with measurable dissolved oxygen levels. The object of biological filtration is to create an environment that supports the largest possible numbers of these bacteria and allows them to metabolize nitrogenous wastes with optimum efficiency. The simplest way to create such conditions is to pass well-aerated, waste-laden water through a porous medium that favours colonization by nitrifiers. It is essential to remember that the active element in a biological filter is its bacterial flora. Experienced aquarists often introduce the appropriate organisms to a new filter bed to accelerate its establishment. Though bacterial starter cultures purporting to serve this purpose are commercially available, their performance to date has not always been reliable. Adding a handful of ceramic rings from a functioning bed to a new canister filter or merely wetting a new sponge filter in an established aquarium works just as well.

Remember, too, that such a filter cannot function at peak efficiency until its bacterial population is fully established. This is a good reason for adding a few fish at a time to a newly set-up aquarium, otherwise the waste load produced by the newly introduced fish will outstrip the growth rate of the bacterial population and lead to an upsurge of toxic byproducts. Marine aquarists have learned to stock a tank initially with one or two organisms that are tolerant of temporarily high levels of ammonia and nitrite. Once the filter is fully established, as indicated by a gradual drop in the nitrite level, additional fishes can be added gradually. This minimizes the likelihood of losing fishes during the filter's running-in period. African cichlid keepers are well advised to adopt this prudent approach, notwithstanding the more robust constitutions of the fishes in their charge.

When stocking a newly set-up cichlid tank in this manner, take care to introduce the least aggressive species first and the most aggressive last. The longer a cichlid occupies a tank, the more its aggressive tendencies will be reinforced. If the most belligerent of a community's intended residents is the first to be added to the tank, it will quickly come to regard it as his own and can be counted on to harass any subsequent additions. Less aggressive species bear up poorly in the face of such persecution. Even if they are not killed outright, the constant behavioural stress to which they are exposed makes them extremely susceptible to systemic bacterial infections.

Finally, never forget that a biologically active filter bed has a finite waste processing capacity, determined by the metabolic rate of its nitrifying population. This in turn is influenced by environmental factors, such as water temperature and dissolved oxygen concentration. The stocking rates suggested here tend to err on the side of caution, so be prepared to experiment cautiously in order to determine the full potential of your system. However, the bottom line is simply this: there is not always room for one more in an aquarium! The temptation to disregard this axiom is pervasive and powerful; the consequences of succumbing to it are inevitable and often painful, as most novice aquarists learn to their cost.

Undergravel filters

The traditional approach to biological filtration – the so-called 'undergravel filter' – uses a tank's substrate as the filter bed. Although undergravel filters are popular with many authorities, their usefulness in a cichlid aquarium is limited. First of all, most African cichlids dig energetically during the breeding period. Such activity exposes an undergravel filter's basal plate, which greatly reduces its effectiveness. Equally important, undergravel filters are immobile and cannot be transferred to another aquarium at times when, for example, fishes are being treated for disease. Such a situation is by no means unusual, but common medications, such as

tetracycline hydrochloride, formaldehyde and complexed copper sulphate may all prove harmful to the essential bacterial flora in a biological filter. The chemicals in these treatments are known either to kill nitrifiers outright or severely inhibit their metabolism.

Happily, many materials other than gravel are more hospitable to colonization by nitrifying bacteria and, since there are no practical constraints on the location of a biologically active filter bed, there are several alternatives to undergravel filtration from which the cichlid fancier can choose.

Sponge filters

Sponge filters, either air driven, or motorized, are an excellent choice for any situation where a minimal waste load is to be treated. The extreme porosity of the sponge material allows it to support a bacterial community comparable to that of an extensive gravel bed. As long as there is no mechanical blockage of the sponge's pores to obstruct the free flow of oxygen-laden water through the sponge medium, such a filter will break down metabolic wastes efficiently.

Such filters are particularly well suited to aquariums for the smaller substrate-spawning species from Lake Tanganyika, as well as for many dwarf and middle-sized riverine cichlids. Air-driven sponge filters are also suitable for nursery tanks for female mouthbrooders, and in fry-rearing tanks. Air-driven units combine efficient operation with complete mobility and are available in a full range of sizes. The smaller units have the further advantage of being relatively unobtrusive and easily camouflaged with rocks or plants.

Internal power filters, incorporating a bioactive sponge medium, are also readily available. These units are particularly useful in larger tanks, where strong water circulation is required. However, their motors generate considerable waste heat and rely on the tank's water for cooling. In this case the filter capacity must closely match the tank volume, otherwise overheating can prove a problem, particularly during the summer. Fortunately, internal power filters are available in a number of different sizes, so it is a simple matter to match the size of the unit to the volume of the tank.

Sponge filters do not perform satisfactorily in an environment where the waste load is so heavy that it actually clogs the pores and obstructs water movement. Such a situation arises in tanks housing very large cichlids, such as the various tilapias, or in those that are very heavily stocked, such as Malawian cichlid communities. In these cases, the most efficient filtration strategy is to combine a high-capacity external power filter with a canister filter.

Canister filters

A canister filter allows the aquarist to move a biologically active filter bed outside the aquarium. Such units are every bit as portable as sponge filters. Furnished with the appropriate medium, such as ceramic or plastic rings, a canister filter can support as large a nitrifying population as an undergravel filter bed. These units cannot be disrupted by the fishes and can include crushed oyster shell or coral gravel if you need to create hard, alkaline water conditions in the tank. Their chief shortcomings are susceptibility to blockage and the inconvenience of servicing them. Using a canister filter in conjunction with an outside power filter greatly minimizes both disadvantages.

Outside motor-driven power filters are simply very efficient mechanical filters. Equip them with an efficient, reusable medium and they will trap particles of waste where these can be easily removed from the system, and before they undergo large-scale biodegradation. In the process, they also trap coarse particles before these can be drawn into a canister unit, where they will eventually obstruct the water flow.

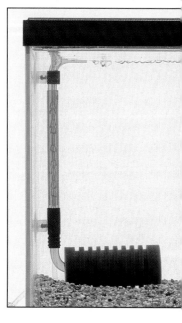

Above: *An air-operated sponge filter is ideal for the cichlid tank with a minimal waste load, and in nursery and fry-rearing tanks.*

Cleaned weekly, an outside filter added to a community of medium-sized to large African cichlids can mean the difference between servicing a canister unit monthly or quarterly. A wide choice of outside power filters is available to aquarists. Remember when making a choice that a unit that cannot be easily cleaned will not be frequently cleaned, and a unit that is not frequently cleaned cannot perform its essential function efficiently.

Given the importance of regular servicing, it is clear that choosing an appropriate filter medium is just as important as selecting the unit itself. Disposable dacron pads clog so quickly in a cichlid aquarium that using them soon becomes uneconomical. Faced with a heavy waste load in the aquarium, it is well worth considering a commercially available plastic filter medium with a weak electrical surface charge that precipitates suspended waste particles on to its strands. This is an extremely efficient medium, notwithstanding a loose texture that maximizes water flow through the filter. It is easy to clean under a stream of warm water and can be reused indefinitely. Although its initial price is considerably greater than that of dacron pads, this material pays for itself many times over within just a few months.

When considering a manufacturer's claims of filter performance, bear in mind the very different functions performed by biological and mechanical units. A mechanical filter operates by pulling particles of material out of suspension. To function effectively, all the water in the tank must pass through the filter medium two to three times per hour. The rate of flow through the filter medium drops dramatically as more waste is trapped, and as a result, power filters never maintain their rated flow under actual operating conditions. For an aquarium with a capacity of 200 litres(44 Imp. gallons/52 US gallons) to work at maximum efficiency, so that the total volume of water in it is filtered twice an hour, it needs a mechanical unit that pumps 800-1000 litres(176-220 Imp. gallons/208-260 US gallons) per hour through an empty filter box.

A high flow rate is less important in a biological filter. For this type of filter to be effective, the waste stream must remain in contact with the bacterial flora long enough for the nitrifiers to work on the dissolved metabolites. It is enough if the water in the tank moves through the filter medium once an hour. If the biological filter becomes blocked, it will cease to work. You cannot compensate for such a blockage by selecting a more powerful pump to drive the filter; instead you must take steps to prevent a blockage by regularly removing particles of waste from the filter bed. If you combine biological filtration with a mechanical filter, you will find that you do not need to clean the filter bed so frequently.

The importance of water changes

In theory it is possible to avoid physiological stress caused by a build up of waste products solely by installing efficient biological filtration in the aquarium. This approach founders on the harsh reality that cichlids are large fishes with healthy appetites and produce a great deal of both liquid and solid waste. Furthermore, many African cichlids exist at high population densities in nature and experience has shown that one way to minimize unwanted aggression is to house these species in a comparable fashion in captivity. This adds further to the waste load generated by a cichlid community. Efficient biological filtration unquestionably plays an important role in nitrogen cycle management in a cichlid aquarium. However, it cannot be expected to do the job unaided and a programme of regular partial water changes is thus an indispensable element of successful cichlid husbandry, regardless of species.

Riverine, Victorian and Malawian cichlids appreciate partial water changes of 30-50 percent every 7-10 days. In a heavily stocked aquarium you will need to increase the frequency of the water changes and the amount of water replaced. Malawian cichlids, in particular, seem to relish a regular replacement of up to 85 percent of their tank's volume. However, Tanganyikan cichlids respond less favourably to large-scale water changes. Representatives of the genus *Tropheus* and its allies appear the most tolerant of such a programme, whereas *Lamprologus* and representatives of allied genera, seem more sensitive to such practices and usually refuse to breed when managed in this way. Fortunately, most of these Tanganyikan cichlids will thrive as single pairs and the waste load produced in such a lightly stocked tank is unlikely to strain the capacity of a well-established biological filter. Wholesale water replacement should be

Right: *An external power filter. Water drawn from just above the gravel bed is cleaned and returned to the aquarium through a horizontal spray bar that runs the length of the tank. Electrically operated units provide an increased water flow compared to air-driven filters. Add an appropriate filter medium (e.g. a sponge block or net bag of ceramic rings). Keep filters clean. In large tanks, combine a power filter with a motorized internal filter for best effect.*

unnecessary; it is enough to change 10-15 percent of the tank's volume every 14-21 days to keep dissolved metabolites at acceptable levels. Changing the same volume of water in a more densely populated Tanganyikan community tank every 7-10 days serves the same purpose without distressing the fishes.

Always ensure that the temperature of freshly drawn water is within 1°C(2°F) of the water in the tank to which it is added. It must also have the same chemical make-up. This condition is clearly easier to meet if your fishes prosper in unmodified tapwater.

As a rule, the chlorine content of most municipal water supplies is not high enough to threaten the well-being of adult African cichlids. Nor is there any evidence that the chlorine concentration in tapwater poses any risk to a biological filter bed. However, it is sensible to use a proprietary dechlorinating agent when making water changes in fry tanks or those containing adults of dwarf species.

Chloramine is quite another story. This compound of ammonia and chlorine is more persistent than elemental chlorine and less

easily neutralized. It is a fairly simple matter to break the bond between the two substances and neutralize the resulting free chlorine with a double or treble dose of a commercial dechlorinating agent. However, there then remains the task of removing the toxic ammonia that remains behind.

The problem of residual ammonia is obviously more important to keepers of Malawian or Tanganyikan cichlids than to those who maintain riverine species. Under the alkaline water conditions demanded by the Rift Lake cichlids, much of the ammonia will exist in the toxic, electrically neutral form (NH_3), whereas under neutral to acidic conditions, the harmless, positively charged ammonium ion (NH_4^+) will predominate. A fully established biological filter can degrade small quantities of ammonia efficiently, so one way of coping with the presence of chloramine is to reduce the volume of water replaced at each change and to increase the frequency of the changes. Alternatively, a proprietary single-step chloramine neutralizing agent, used according to the manufacturer's instructions will tie up the ammonia in a harmless organic complex that can subsequently be degraded by bacterial activity.

Aquarists who live in areas with a severe winter climate may have to take into account the risk of gas embolism when making water changes during cold weather. Sometimes known as gas-bubble disease, the condition is caused by the supersaturation of dissolved gases in very cold water. As the water warms up, it loses its capacity to hold the dissolved gas. The result is an eruption of bubbles over every solid surface that the dissolved gas contacts as it leaves the water. When such bubbles form in the superficial tissues of a fish's skin, or in its gill filaments, they can cause serious damage. In small fishes it can often prove immediately fatal; in those species large enough to survive the initial trauma it can lead to secondary bacterial infections.

To test for gas supersaturation draw a glass of cold tapwater and let it sit at room temperature for fifteen to twenty minutes. The presence of air bubbles covering the interior of the glass indicates the presence of dangerous concentrations of dissolved gas in the water supply. The only way to completely eliminate the risk this condition poses to the fishes, is to draw water a day before making a water change and allow the dissolved gases to effervesce as they reach room temperature. However, this approach is not always practical for aquarists, since it entails storing a considerable quantity of water. Refilling a tank through a mist nozzle as used by gardeners will dissipate most dissolved gases before they enter the tank, but this is a time consuming process, as the flow of water through such a nozzle is very slow. A more realistic alternative is to restrict water changes to 10 percent of a tank's volume during the winter months and compensate by increasing their frequency.

Remember to disconnect the tank's heater when making a water change. If the water level drops too far, the heater tube will seriously overheat and almost certainly crack when it comes into sudden contact with cooler replacement water. Many experienced aquarists also believe that it is not advisable to combine large-scale water changes with the replacement or cleaning of the tank's filter medium. Insofar as the presence of organic matter tends to neutralize free chlorine, there is some truth in this as far as a mechanical unit is concerned. However, the function of a biological filter will not be affected if the medium is rinsed in warm, rather than hot, tapwater.

Most of the mess and bother associated with replacing water are eliminated if one uses an automatic water changer. This apparatus – part siphon-powered gravel cleaner, part water bed pump – attaches directly to the tap and, with a flip of a control knob, changes from a powered siphon to a hose. Thus you can combine a water change with the removal of waste from the substrate, another important aspect of nitrogen cycle management.

Breeders may find the action of an automatic water changer a bit too powerful in a tank containing newly mobile fry. It is a good idea to collect the waste water from a fry tank in a bucket before pouring it away, thus allowing any wayward fishes to be rescued and returned to their tank. A close encounter with an automatic water changer would result in a free one-way tour of the local waste treatment plant. However, there should be no problem finding an alternative means of cleaning a fry tank. The many aids available to the fishkeeper today have transformed the nature of time-consuming, often tedious, but essential chores. The efficiency with which you can carry out routine maintenance, will in turn increase your success in the hobby and contribute to your enjoyment of it.

Heating and lighting

The majority of African cichlids are fish of the lowland tropics. While many riverine species regularly encounter temperatures as high as 35°C(95°F) in nature, no cichlid found in such habitats ever experiences temperatures lower than 18°C(64°F). Rivers flowing through open savanna may experience daily temperature fluctuations of up to 7°C(13°F) during the dry season, but the water temperature of streams flowing under forest cover may vary by no more than 1-2°C(2-5°F) over a 24-hour period. Inshore habitats in the Great Lakes are virtually devoid of significant diurnal temperature variation and vary by no more than 1-2°C(2-5°F) on a seasonal basis.

In captivity, African cichlids prosper over a temperature range of 21-29°C(70-85°F). Most aquarists prefer to maintain them at the upper end of this range, on the grounds that spawning occurs more readily when these fish are kept warmer. In the case of mouthbrooding species an added benefit is that the incubation period is shorter at higher temperatures. On the other hand, cichlids are significantly more aggressive at the upper end of their preferred temperature range. Their metabolism is also accelerated, leading to heftier appetites and an increased waste load for the aquarist to cope with. There is also reason to believe that the lifespan of dwarf forest-dwelling riverine cichlids and of the popular mbuna of Lake Malawi are shortened by continual exposure to higher temperatures. Generally speaking, however, the advantages of maintaining African cichlids between 21-23°C(70-75°F) outweigh any disadvantages. To trigger spawning increase the temperature by a few degrees.

Cichlids are able to tolerate changes of 1-2°C(2-5°F) in either direction over a period of 45 minutes to an hour. To some extent this simplifies making water changes, but more extreme changes may provoke an outbreak of 'ich'. Tanganyikan cichlids, in particular, are extremely sensitive to abrupt temperature fluctuations, particularly downwards. A reliable thermometer is an essential piece of equipment but unfortunately, precision is not a feature of the thermometers manufactured for aquarium use. To overcome this problem, invest in an accurate darkroom thermometer, and use it to callibrate the tank thermometer.

A reliable, thermostatically controlled heater is obviously vital. Units with submersible heater elements are particularly efficient in large tanks, where they can be sited to take advantage of warm water's tendency to rise. This greatly simplifies the task of ensuring that the aquarium is uniformly heated. However, any well designed and constructed unit can be successfully incorporated into an African cichlid tank.

Heaters that use solid-state thermostats, although more expensive, are far less vulnerable to failure than are units that rely on old-fashioned bimetallic strip thermostats. Always match the heater output carefully to the actual tank volume. Calculate the volume of water in the tank by subtracting from the rated tank volume the space taken up by the gravel and furnishings. Alternatively, keep a note of the quantity of water used to fill the tank for the first time, after setting it up. A heater's output is measured in watts. The higher the rated wattage of a unit, the greater its output of heat. Allow 10 watts/4

litres (approximately a gallon) of actual tank volume. Limiting the heater's wattage to a sensible maximum guarantees that the water in the tank will never be heated beyond a temperature that may prove lethal to the stock.

Lighting

Unless you plan to combine aquatic horticulture with cichlid husbandry, you need not worry unduly about lighting. The usual fluorescent tubes serve this purpose quite satisfactorily. Most African cichlids are indifferent to light intensity and can be expected to do well in brightly lit tanks. The smaller forest-dwelling riverine species and a number of cichlids found at depths in excess of 3m(10ft) in the Great Lakes prefer more diffuse illumination in captivity. A screen of floating plants on the surface of their tanks is enough to satisfy them.

Tungsten or metal halide spotlights have serious practical disadvantages as a light source for the aquarium since they generate a tremendous amount of heat and must be mounted 45-50cm (18-20in) above the top of the tank. It is therefore impossible to incorporate these lights into standard aquarium lighting hoods and many aquarists find it difficult to devise alternative mounting arrangements. Nor is it possible to use them in conjunction with glass covers. Even placed well above the water surface, they generate sufficient heat for the so-called 'greenhouse effect' to operate.

Above: *African cichlids prosper in bright or dim lighting conditions. For plant growth, combine warm white and colour enhancing tubes.*

This seriously complicates the task of maintaining a constant tank temperature. Secondly, because of their high wattages, spotlights are costly to operate and often impose such a strain on older electrical systems that the outlets must be rewired to accommodate them. Lastly, they are expensive and have short working lives, which entails their frequent replacement.

If you require vigorous plant growth in the African cichlid tank, illuminate it at the rate of 1 watt per 4 litres (approximately a gallon) of water, using either a combination of warm white and colour enhancing bulbs or full spectrum fluorescent tubes, for 14 hours per day. This formula assumes a tank depth of 30cm(12in). For every additional 8.0cm(3.2in) of tank depth, double the wattage value in the above formula. As rated wattages increase in 5- and 10-watt increments, this is only a guide and it is not always possible to install a lighting system that corresponds exactly to the value of the calculations. In such instances, always round the actual wattage of the tubes upwards to the closest possible value. If excessive algal growth should occur, gradually cut back the number of hours that the tank is lit each day. Often the only way to bring an algal bloom under control is to reduce the number of fish in the tank.

Aquarium selection and aquascaping

Given the group's great diversity, there is little point in trying to lay down detailed rules on what constitutes a suitable aquarium for African cichlids as a whole. Specific recommendations are given under each major subgroup's heading in the species section of this book. However, it is always worth investing in the largest aquarium you can accommodate. The undesirable behaviour for which the family is notorious in captivity is much less obvious when cichlids have plenty of living space. Cichlids are also rather messy fish; their waste output poses a real challenge to successful nitrogen cycle management (see page 23). The larger the tank, the easier it is to keep dissolved nitrogenous wastes at acceptable levels.

Selecting a tank

Because most cichlids live in fairly close association with the bottom, it pays to select relatively shallow tanks with an extensive base area, rather than taller, so-called 'show tanks'. An exception to this rule are tanks for the mbuna of Lake Malawi; under aquarium conditions they will happily swim among vertical rock faces. A handful of pelagic species endemic to Lake Tanganyika, such as the several *Cyprichromis* species and the featherfins of the genera *Cyathopharynx* and *Ophthalmotilapia*, also appear to appreciate deeper tanks. For the most part, however, cichlids make little use of the upper two-thirds of the water column in their tank.

Regardless of size, an African cichlid aquarium requires a tight fitting cover. Do not buy a tank that lacks a secure cover unless you can modify the design and add a cover later. Cichlids, unlike some other fishes, do not jump in pursuit of food, or with the object of migrating towards some predetermined destination. However, a cichlid on the losing end of a serious quarrel may well resort to a frantic leap up and out of its tank. This tactic may allow a defeated fish to escape further harm in nature, but in captivity few cichlids benefit from a prolonged sojourn on a rug or concrete floor. The knowledge that such a loss is so easily prevented merely serves to heighten the distress an aquarist feels when he discovers the desiccated remains of a prized missing fish.

African cichlids, with few exceptions, require large tanks if they are to prosper. It is only sensible, therefore, to devote serious thought to how such an aquarium is to be supported. Remember that water weighs one kilogram per litre (approximately 10lb/Imp. gallon; 8.4lb/US gallon). Allowing for its own weight, a 200 litre(44 Imp. gallon/53 US gallon) aquarium – the smallest recommended quarters for a community of Lake Malawi cichlids – represents just under 205kg(approx. 450lbs) of concentrated weight filled with water alone. Few pieces of furniture can support such massive objects. In addition, there is always the small, but real, possibility of furniture being damaged by a leak or by the spilling and splashing of water that is part and parcel of routine tank maintenance. Buying a commercial stand for any aquarium in excess of 80 litres(18 Imp. gallons/21 US gallons) capacity is therefore the most practical solution to this problem. Metal or wood aquarium stands are available in a wide selection of styles, so it is quite easy to find one to complement the decor of any room in the house.

Above: *In a Tanganyikan cichlid tank, boulders and rocks create separate territories and provide spawning sites. Plants add interest and variety.*

Right: *In this plan, good room positions are shown in blue, reasonable sites are in purple, unsuitable ones in pink.*

May be too light/difficult to decorate

Natural daylight

Tank may receive too much sunlight

Keep tank away from door

Siting the tank

African cichlids produce copious quantities of nitrogenous wastes, so the tanks that house them are more susceptible than most aquariums to explosive algal growth. Many African cichlids have marked herbivorous tendencies, so relying on aquarium plants to soak up this abundance of fertilizer is not always possible. Try to locate the tank where you can control the lighting. It is much easier to manage algal growth if the tank is placed away from any source of natural light. Failing this, north – or east – facing windows are acceptable locations in the Northern Hemisphere. Tanks facing south or west are a standing invitation to unsightly algal blooms.

Finally, it is simple common sense to position a tank as close as possible to an electrical outlet. Avoid trailing extension leads, both for aesthetic and safety reasons. At one time, it was important to site the aquarium close to a drain to simplify routine maintenance. Today, the availability of automatic water changing and/or gravel cleaning devices makes it no longer essential, although undeniably convenient, to locate the tank near a water supply.

Aquascaping

The primary reason for aquascaping an aquarium is aesthetic. Fishes in general, and African cichlids in particular, will live and breed happily in aquarium surroundings that bear very little resemblance to their native habitat. However, to the observer, ornamental fishes unquestionably make a more satisfactory showing in a reasonably naturalistic setting, with the proviso that the biotopes of most African cichlids cannot be duplicated exactly in captivity. Nor, in the majority of instances, would aquarists find the result of such an exercise to their liking if they could. Aquascaping an African cichlid tank is like gardening; its object is to create a setting that shows the fish to best advantage, while satisfying their needs for shelter and breeding sites.

This task is complicated by the diversity of habitats occupied by these fishes. The different water requirements of riverine and Rift

Lake cichlids limit the appropriate materials that you can use to aquascape their quarters. African cichlids also differ in their need for cover and in their tolerance of live plants. It is thus impossible to offer a single set of recommendations for aquascaping an African cichlid aquarium. Instead, it is more practical to consider the various substrate materials and furnishings appropriate to the major groups of African cichlids, and then make some specific suggestions on setting up an aquarium to accommodate them. Do remember that unless otherwise stated, recommended tank sizes are based on the assumption that you are housing a single breeding group per tank. If you wish to keep African cichlids in a community setting, add 50 percent of the minimum recommended tank area to the base figure for each additional pair or trio to be housed in the aquarium.

Selecting a suitable substrate
When you come to select the aquarium substrate, remember to take into account its chemical composition, since this will in turn affect the water chemistry in the tank. Any material intended for use in a tank housing soft, acid-water species must be either chemically inert or have an acidic reaction, lest it raise pH and hardness values to unacceptable levels. Pure silica substrata, such as cage bird grit or fine silica sand, are chemically inert. So are quartz and flint gravels and epoxy coated materials. Laterite and many basalt-derived gravels, on the other hand, will slightly acidify aquarium water.

To test for soluble minerals capable of raising pH and hardness, place a small sample of the substrate on a glass or plastic plate and add a few drops of dilute hydrochloric acid or even vinegar. If the material effervesces, it contains significant quantities of soluble carbonates. Such a substrate will eventually harden the water of any aquarium to which it

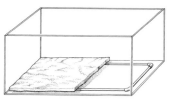

Above: *As a change from sand and gravel substrates, provide a piece of flagstone resting on a PVC pipe base. This novel form of tank decor is ideally suited to mbuna, the rockfish of Lake Malawi.*

is added and is not suitable for use with species that require soft, acid water to prosper.

In most instances, the chemical composition of the tank's substrate is a matter of indifference to riverine cichlids, while those native to Lakes Tanganyika and Malawi can only benefit from the use of materials rich in soluble carbonates on the tank base. Ordinary washed river or beach gravel suits the needs of the first group quite well. Aquarists living in areas with extremely soft tapwater often find it worth the extra expense of obtaining coral gravel as a substrate for Rift Lake cichlid tanks because of the extra buffering action it provides. Crushed oyster shell is not suitable for this purpose; it becomes too tightly packed, creating 'dead' pockets where anaerobic decomposition can occur. When disturbed, these can release toxic hydrogen sulphide into the aquarium, often with disagreeable consequences. You can also use crushed dolomite as a substrate in such an aquarium, but its suitability as a water conditioning agent is questionable, as noted on page 22.

Unless you intend to aquascape an African cichlid tank with live plants, do not use more of any substrate material than you need to cover the base of the tank to a depth of 1.25-2.5cm(0.5-1in). This sets the fish at ease by providing a non-reflective bottom, yet minimizes the opportunity for waste to accumulate.

As an alternative to traditional substrate materials, you could consider resting a single piece of flagstone, cut slightly smaller than the internal dimensions of the tank, on a framework of PVC pipes. This arrangement allows water to circulate freely under the slab, while offering the tank's less assertive inhabitants a refuge of last resort when harassed by more aggressive companions. It is particularly appropriate for an aquarium housing mbuna or their Tanganyikan counterparts, since most of these cichlids seldom encounter sand or gravel in nature.

Using live plants

The contemporary growth of interest in the culture of live aquatic plants is, on balance, a positive trend in aquarium keeping. Regrettably, it has been accompanied by a tendency to regard the unplanted aquarium as an 'unnatural' environment, in which fishes are unlikely to prosper. Such an attitude is certainly mistaken as far as African fishes in general, and African cichlids in particular, are concerned. The fresh waters of Africa are not richly endowed with aquatic plants, while biotopes dominated by them are very rare. The extreme seasonality of most savanna rivers imposes severe constraints on rooted aquatic plants. Only those species capable of surviving the dry season, such as tubers, bulbs or seeds, prosper

under these conditions. Less hardy plants will thrive only in the relatively rare permanent marshes and oxbow lakes found in the lower courses of these rivers.

Although forest streams do not exhibit such a seasonal pattern of flow, they are similarly lacking in rooted plants because their currents are often swift, their waters are poor in nutrients and often heavily shaded by the overhanging canopy.

Given this state of affairs, it should be no surprise that while many African cichlids will live happily in a well-planted aquarium, a profusion of rooted plants is hardly essential to their well-being. Most species, however, appreciate a layer of floating plants in their tank. Such cover affords shy species a sense of security and goes a long way towards bringing them out into the open, where they are easier to observe. However, more elaborate attempts at aquascaping are hardly essential. Even forest-dwelling dwarf species, such as the several *Nanochromis* and *Pelvicachromis* species, will breed freely in a tank containing nothing more than a layer of water sprite (*Ceratopteris pteridoides*) at the surface and a few clumps of Java moss (*Vesicularia dubyana*) and Java fern (*Microsorium pteropus*) scattered among the furnishings.

Another factor to consider when choosing live plants for an African cichlid tank is the tendency of many species to treat such aquascaping as a self-service salad bar. All tilapias, for example, are herbivorous to a greater or lesser degree. Even species that do not normally feed on aquatic plants in nature will browse on them in captivity. With the exception of the several *Labidochromis* species and the long-snouted, predatory *Melanochromis* species, so will the colourful mbuna of Lake Malawi and their Tanganyikan counterparts of the genera *Petrochromis*, *Simochromis* and *Tropheus*.

Below: *Since this substrate-spawning* Chalinochromis marlieri, *will dig at spawning time, be sure to plant* Anubias *in containers.*

Plants suitable for African cichlid tanks

Anubias barteri
Do not bury rhizome. Prefers moderate light. Herbivores find tough leaves distasteful. Hardy but slow growing.

A. nana
Dwarf anubias
Smaller than *A. barteri* but identical in requirements and growth pattern.

Aponogeton crispus
Wavy-edged swordplant
Provide bright light. Easily grown from bulbs. Tolerates a wide range of hardness and pH values.

A. ulvaceus
More likely than *A. crispus* to become dormant and shed leaves for a time. Otherwise differs only in leaf shape.

Bolbitis heudelotii
African fern
Do not bury rhizome. Prefers moving water and dimly lit tank. Distasteful to herbivores. Slow grower.

Ceratophyllum demersum
Hornwort
Provide bright light and thin out regularly.

Ceratopteris cornuta
African water sprite
Needs bright light. Easy to grow. Thin out frequently. Grows equally well floating or planted.

C. pteridioides
Water sprite; floating fern
As for *C. cornuta*. Does not thrive as a rooted plant.

C. thalictroides
Fine-leaved water sprite; Indian fern
As for *C. cornuta*. Does best planted, rather than floating.

Crinum thaianum
Thai onion plant
Plant neck of bulb above substrate. Leaves grow up to 1.5m (5ft) long.

Cryptocoryne wendtii
Prefers a moderate light. Tolerates hard, alkaline water better than most cryptocorynes.

C. willisii
Smaller than *C. wendti*, but also tolerates hard, alkaline water. Both species relish frequent water changes; avoid prolonged exposure to elevated nitrate levels.

Echinodorus major
Ruffled swordplant
Provide bright light; Easy to grow in large (120 litre/30 gallon) tank.

E. parviflorus
Black Amazon swordplant
Easiest of the medium-sized swordplants to grow. All need bright light and tolerate high nitrate levels better than *Cryptocoryne* spp.

Hygrophila difformis
Water wisteria
Needs bright light. Pinch back emergent shoots to prevent loss of finely-divided submerged foliage.

Microsorium pteropus
Java fern
Do not bury rhizome. Does best in dimly lit tanks. Distasteful to herbivores. Slow grower.

Nymphaea lotus
Red or tiger lotus
Needs bright light. Remove floating leaves to encourage vigorous growth of submerged leaves. Solid red, and maroon-speckled green leaved varieties also available.

Vesicularia dubyana
Java moss
Non-rooted. Prefers dim light.

Decorative materials suitable for African cichlid tanks

Material	Advantages	Disadvantages
Driftwood Wide range of sizes available commercially. Do not collect pieces from local streams or seashore – they may contaminate the tank.	Light, easy to work with; natural appearance. Carve out hollows to accelerate waterlogging and provide added shelter.	Buoyant; secure until waterlogged. May lower pH level of water.
Tropical woods e.g. mangrove	Denser than water so they drop immediately to bottom of tank. Suitable for West African forest-dwelling cichlids.	May discolour and acidify water. Soak for a week in 15 percent solution of bleach, replaced daily, then rinse thoroughly. Not for Lake cichlid tank.
Basalt, tuffa, sandstone	Light; easy to work with	Not always available
Limestone	Eroded stone is lighter and provides cover. Calcareous; suitable for Rift Lake tank. Provides additional buffering capacity.	Heavy; occupies substantial volume in tank.
Shale, flagstone, slate	Make good caves and overhangs. Chemically neutral. Suitable for forest and Rift Lake cichlids.	
Plastic and ceramic e.g. imitations of rockwork/driftwood.	Readily accepted by fishes. Easy to acquire. Do not look too artificial.	Providing a sufficient quantity can prove expensive.
Snail shells	Calcareous; safe for Rift Lake cichlids. Many lamprologine sp. rely on them for shelter and spawning sites. Shells for marine tanks also suitable for cichlids.	Not suitable for soft water species.
Miscellaneous Clay flowerpots, empty coconut shells; sections of PVC pipe.	Light and portable. Easy to modify for tank use.	Less decorative than natural materials.

Finally, the spawning behaviour of some African cichlids complicates the culture of rooted plants in their tank. These species are not herbivorous, but many of them move tremendous volumes of gravel at the onset of the breeding period, either to construct a spawning pit or to provide shelters for newly hatched fry. The plants described on page 36 are not all of African origin but they can be expected to prosper in an African cichlid tank with a minimum of care.

Aquarium infrastructure

African cichlids require a substantial infrastructure in their tanks. These structures break up an aquarium into a series of more or less separate zones. Each family of fishes thus acquires a focal point for its characteristic territorial behaviour, and the tank is able to support a greater number of cichlids. Suitable materials, carefully selected and positioned in the tank, provide shelter and create a visually pleasing setting. Indeed, where live plants are impractical, aquascaping is necessarily limited to selecting and arranging rockwork, driftwood and other inanimate materials (see page 37). The disadvantage of such an arrangement is that it may be difficult to catch the tank's residents. This explains why breeders, who are less concerned with aquarium aesthetics, prefer to use more manageable materials, such as flowerpots and sections of PVC pipe to provide their cichlids with adequate cover.

Tankmates for African cichlids

With the exception of certain predators, such as the banded jewel fishes, *Cyphotilapia frontosa*, the larger *Lamprologus* species and a number of haplochromine cichlids native to the Great Lakes, sexually quiescent African cichlids will usually ignore non-cichlid tankmates that are a third to one half their overall length, or even larger. Since cichlids prefer the lower third of the water column,

midwater-swimming fishes fill a visual gap in a tank. The untroubled behaviour of schooling species also adds greatly to shy cichlids' sense of security; indeed, the judicious addition of such dither fish to their aquarium is a well established technique for bringing the most retiring dwarf cichlids out into the open.

Suitable dither fish must be too large to tempt the appetites of their cichlid neighbours, yet not so large that they offer serious competition at feeding time or pose an unreasonable threat to cichlid fry. Small to medium-sized danios, barbs or rasboras make good dither fish for forest-associated cichlids, since they prefer the same water chemistry. The smaller labyrinth fishes also make good tankmates for West African dwarf cichlids. Poeciliids and Australasian rainbowfishes are at home in the hard, alkaline water demanded by Lake Tanganyikan cichlids and make suitable companions for all but the most predatory cichlids. Goodeids also find such water conditions to their liking, but be cautious when housing them with small cichlids. Some of the more robust goodeids are so aggressive that they function as anti-dither fish, terrorizing their cichlid neighbours so badly that they refuse to leave their hiding places any longer than is necessary to snatch a mouthful of food!

Few traditional scavenger fishes make good tankmates for African cichlids. Loaches of the genera *Botia* and *Noemacheilus* and the 'sharks' of the genus *Labeo* have the same requirements for shelter as small and medium-sized cichlids, and generally assert their rights forcefully, often with disagreeable results for the cichlids. *Corydoras* and other mailed catfishes are a target for cichlid harassment and do not do well, even alongside the smaller African cichlids. This vulnerability extends to most small smooth-skinned catfishes, although in nature their larger representatives,

Above: *This giant South American loricariid catfish* Pterygoplichthys *is a suitable tankmate for larger African cichlids. Select suitable size tankmates for smaller cichlids.*

in turn, often become significant cichlid predators!

The two catfish groups most likely to prosper in an African cichlid tank are the upside-down catfishes (Family Mochokidae) and the armoured suckermouth catfishes (Family Loricariidae).

The most commonly available upside-down catfishes are *Synodontis* species, although representatives of the genera *Brachysynodontis, Hemisynodontis* and *Mochokiella* are sometimes found in importers' tanks. Upside-down catfishes also require access to shelter, but seem better able to coexist with African cichlids than many other territorial bottom-dwellers. They will clean up uneaten food from the bottom and, if hungry enough, many species will graze on algae as well. However, their chief role in a cichlid tank is to pose a potential threat to eggs or fry. In so doing, they provide a focus for the aggression that a breeding pair would otherwise direct towards one another. Obviously, you cannot choose upside-down catfish for the role of 'target fish' if they are much larger than the parental cichlids, otherwise they may overwhelm the cichlid defences. Apart from this, their willingness and ability to defend themselves and marked regenerative abilities make them good target fish for those cichlids that practice prolonged custodial care of their fry.

Loricariid catfishes make ideal scavengers in an African cichlid tank. They can be counted on to clean up uneaten food promptly and, if kept a bit hungry, will do a good job of controlling algal growth on solid surfaces. As with upside-down catfishes, it is a good idea to select a loricariid of a similar size to the breeding cichlids in the tank. Slender-bodied species (*Farlowella* and *Rineloricaria* spp.) and such dwarfs as *Otocinclus* and *Parotocinclus* are too easily damaged to be risked with even the smallest African cichlids. However, this still leaves a wide selection of totally suitable loricariids, from the diminutive *Peckoltia* and medium-sized *Ancistrus* to giants, such as *Hypostomus, Panaque* and *Pterygoplichthys*. In any event, the behaviour of different species of both mochokid and loricariid catfishes varies considerably and it is always a sensible precaution to consult an appropriate reference before adding a given species to an African cichlid tank.

Recommended layouts

Let us now examine in more detail the various aquascapes suitable for different groups of African cichlids. The information in the following tables will help you to select and furnish a suitable aquarium for each group and gives useful guidelines on filtration and stocking levels.

Aquascaping for the African cichlid tank

Type of cichlid	Tank size No. of fishes	Filtration
Small substrate spawning		
Anomalochromis thomasi *Thysia ansorgii* *Pelvicachromis* spp. *Nanochromis* spp. Dwarf *Julidochromis* spp. or *Telmatochromis* spp. Ostracophil *Lamprologus*	60×30cm (24×12in) to 75×30cm (30×12in) 6 sub-adults **OR** 1 adult pair of any monogamous species, plus a few dither fish and 1 medium-sized loricariid catfish **OR** a harem of 1 male and 2 or 3 females of any ostracophil *Lamprologus* species.	Sponge filter or small motorized outside unit with biological sponge block or a bag of ceramic rings.
Medium-sized monogamous		
Chaytoria joka *Chromidotilapia* spp. *Steatocranus* spp. Large *Telmatochromis* spp. Large *Julidochromis* spp. Smaller *Hemichromis* spp. Medium-sized *Lamprologus* spp. *Eretmodus cyanostictus* *Spathodus* spp. *Tanganicodus irsacae*	90×30cm (36×12in) to 120×45cm (48×18in) 6 sub-adult **OR** 1 medium-sized adult pair plus dither fish and 1 medium-sized loricariid.	Inside sponge filter plus large outside power unit.
Large monogamous		
(Adult size over 20cm/8in) Large *Hemichromis* spp. *Tilapia* spp. Large *Lamprologus* spp. *Boulengerochromis microlepis* *Sarotherodon* spp.	120×45cm (48in×18in) 1 pair adults plus dither fish plus 1 or 2 catfish **OR** 150×45cm (60×18in) to 180×50cm (72×20in) 1 large adult pair **OR** 3 or 4 smaller adult fish.	Canister plus large outside power filter.

Aquascaping

General

Substrate: Fine silica gravel 2.5cm (1in deep).
Infrastructure: Low-ceilinged caves, coconut shells, flowerpots. Empty snail shells for *Lamprologus* sp.
Plants: Java and African fern on surface of substrate or fastened to rock or driftwood with nylon thread. *Anubias* spp. in bulb pans, set in gravel and concealed behind rock or wood. Java moss and floating plants.

More likely to breed if offered wide choice of breeding sites. Remove dither fishes and/or catfishes after cichlids have spawned. These cichlids dig in conjunction with spawning but not sufficiently to uproot most plants in the aquarium.

Substrate: Medium gravel 2.5-5cm (1-2in) deep.
Infrastructure: Low-ceilinged caves, coconut shells, flowerpots arranged to create separate territories. Plants are not essential: driftwood and rockwork structures will suffice.
Plants: Java fern, African fern. Live plants in pots – protect surface of pot with large pebbles or pieces of broken flowerpot to discourage digging. Floating plants. Plastic plants secured to flat rocks with nylon thread.

For each additional pair, increase the minimum base area of tank by 50 percent. Remove dither and catfish after spawning. These cichlids do not eat plants, but they dig extensively.

Substrate: Shallow layer of medium to coarse gravel.
Infrastructure: Large single pieces of rock or driftwood, coconut shells, large flowerpots, sections of PVC pipes arranged to create separate territories.
Plants: Artificial plants secured with nylon thread to flat rocks. Floating plants in all tanks except those housing *Tilapia* spp.

ALL ROCKWORK MUST REST SECURELY ON THE TANK BOTTOM. These strong cichlids have no difficulty rearranging the interior decor of their tank, with potentially dangerous consequences – a falling rock can crack a glass panel. All species dig extensively when breeding, while *Tilapia* spp. eat live plants in the aquarium.

Aquascaping for the African cichlid tank

Type of cichlid	Tank size No. of fish	Filtration
Mbuna and Tanganyikan counterparts *Labidochromis* spp. *Pseudotropheus minutus* *Iodotropheus sprengerae*	90×30cm (36×12in) 1 male and 3-4 females.	Large motorized internal filter plus medium-sized outside power filter.
Medium-sized *Pseudotropheus* spp. *Labeotropheus* spp. *Cyathochromis* spp. *Gephyrochromis* spp. Medium-sized *Melanochromis* spp. *Tropheus* spp.	150×45cm (60×18in) 1 male plus 3-5 females of a single species or up to 3 pairs of different spp. **OR** 180×45cm (72×18in) Up to 5 pairs or trios of small to medium-sized mbuna.	Canister filter plus large outside power filter.
Large *Pseudotropheus* spp. *Melanochromis* spp. *Petrotilapia* spp. *Petrochromis* spp. *Simochromis* spp.	180×45cm (72×18in) 1 male and 3-6 females of a single species **OR** up to 3 trios of different species.	Canister filter plus large outside power filter.
Medium-sized open bottom Malawian, Tanganyikan and Victorian and riverine mouthbrooders. Sandy bottom-dwelling Malawian *Haplochromis* (chisawasawa). Pelagic zooplankton-feeding Malawian *Haplochromis* (utaka). *Aulonacara* spp.; *Callochromis* spp.; *Cyprichromis* spp.; small to medium-sized Lake Victoria and riverine *Haplochromis* spp. and related genera; *Pseudocrenilabrus* spp.	120×30cm (48×12in) Single species group of 1 male and 3-4 females **OR** 180×45cm (72×18in) 3 to 5 trios of different species, depending on adult size.	Large motorized internal filter and large outside power filter.
Large riverine and lake mouthbrooders *Aristochromis christyi* *Chilotilapia rhoadesi* Large predatory Malawian *Haplochromis* spp. *Ramphochromis* spp. *Cyphotilapia frontosa* *Cunningtonia longiventralis* *Cyathopharynx furcifer* *Ophthalmotilapia* spp. *Oreochromis* spp. Large Lake Victoria *Haplochromis* and allied genera.	180×45cm (72×18in) Single species group of 1 male and 3-5 females **OR** 2 to 3 trios of different species, depending on adult size.	Canister filter and large outside power filter.

Aquascaping	General
Substrate: 2.5-5cm (1-2in) deep. Use calcareous materials in soft water areas or washed river/beach gravel in hard water areas. **Infrastructure:** Tuffa and highly eroded limestone or flowerpots and sections of PVC pipe joined together to provide cover. **Plants:** Giant Sagittaria, *Crinum thaianum*, larger *Anubias* spp. Java fern fastened with fine nylon thread to rockwork. Floating plants. **Aquarium design:** Place pots along rear wall of tank, concealed behind low rock formations. Stack medium-sized rocks along rear half of each end panel right up to water surface to create ample cover and large common area.	House these polygamous cichlids in single species, one male/multiple female groups, or as pairs/trios in well-stocked community tank. Hyperactive, highly aggressive fish – provide the largest possible tank. Very sensitive to nitrogen cycle mismanagement. Herbivorous – use only recommended rooted plants; will eat hornwort, duckweed and *Salvinia*.
Substrate: No more than 5cm (2in) deep. Use calcareous materials in soft water areas, washed river gravel in hard water areas. **Infrastructure:** Tuffa or eroded limestone. Alternatively, flowerpots or sections of PVC pipe. **Plants:** *Echinodorus* spp.; giant *Sagittaria, Crinum thaianum*, larger *Anubias* spp. Java fern fastened to rockwork with fine nylon thread. **Aquarium design:** Use infrastructure to break tank bottom into several large open areas. These fish require less cover than the rock-dwelling spp.	All have a polygamous mating system – do not house as isolated breeding pairs. Very sensitive to nitrogen cycle mismanagement.
Substrate: Shallow, to 5cm (2in) deep. Use calcareous materials in soft water areas, washed river gravel in hard water areas. **Infrastructure:** Tuffa or eroded limestone for Malawian and Tanganyikan species, driftwood for riverine and Victorian species **OR** flowerpots and PVC pipe. **Plants:** Live plants not recommended. Tilapias are herbivorous, while other species dig extensively when breeding. Floating plants ignored. **Aquarium design:** Leave large open areas for males to use as spawning territories.	All have polygamous mating systems – do not house as isolated breeding pairs. Malawian and Tanganyikan species very sensitive to nitrogen cycle mismanagement.

Feeding

Cichlids fall into one of four subgroups in terms of their dietary requirements: omnivores, micropredators, piscivores and herbivores. You must satisfy the particular nutritional requirements of each group if the fishes are to thrive in captivity. Fortunately for today's aquarist, the wide selection of good-quality prepared and frozen foods available commercially makes it a simple matter to provide any African cichlid with a palatable and nutritious diet. It is unwise to include *Tubifex* worms in the diet of any African cichlids. There is a strong correlation between their frequent appearance on the menu and the incidence of systemic bacterial diseases in cichlids.

It is worth noting that all cichlids tend to gluttony in captivity, although such behaviour is seldom a reflection of their nutritional state. In nature, many African cichlids rely significantly on high-roughage foods, such as blue-green algae or organic detritus. Although often abundant, such resources supply little food value per gram and in order to meet its metabolic needs the fish must eat continuously. Not surprisingly, African cichlids with this feeding pattern attempt to do just the same in captivity, regardless of the greater nutritional value of the foods offered to them. At the other extreme, large predators exploit relatively scarce prey items of high nutritional value. In the wild they seldom, if ever, have the opportunity to eat to satiety, so natural selection has never operated to place a limit on their feeding behaviour. Hence, these fish will also overeat whenever afforded the chance.

It is obviously essential to offer any captive animal sufficient food to enable it to grow normally, remain healthy and, if the opportunity arises, to breed successfully. However, you need not feed to excess to accomplish these ends. Overfeeding high-protein, low-roughage foods may actually harm a cichlid's health and will inevitably make nitrogen cycle management in the aquarium more difficult. To ensure the well-being of African cichlids, offer them a diet based on their known nutritional requirements and stick to a schedule of regular feedings. As a useful rule of thumb, offer African cichlids no more food at any one time than they can consume in five minutes.

Do not hesitate to include a weekly fast day in this regime. One day in seven without food is hardly unnatural for any fish and allows captive specimens to purge their intestinal tracts on a regular basis. Equally important, it eases the burden on a tank's biological filter, thus simplifying nitrogen cycle management to a degree.

Regardless of their feeding patterns, many African cichlids need a regular supply of foods that contain beta-carotene and canthaxanthin to retain the full intensity of their coloration. Fresh vegetable foods and frozen zooplankton are good sources of both these substances. Alternatively, you can add one of the commercially available colour enhancing foods to the fish's normal diet. This is the easiest way of ensuring that captive fishes receive a regular supply.

Below: *Even herbivorous cichlids require a weekly feeding of animal protein. Here,* Tropheus *sp. relish an offering of red mosquito larvae.*

Feeding African cichlids

Feeding group	Recommended diet
Omnivores *Anomalochromis thomasi*, *Chaytoria joka*, *Chromidotilapia* spp., most small *Haplochromis* spp., *Melanochromis johanni*, *Nanochromis parilius*, *Oreochromis* spp., *Pseudocrenilabrus multicolor*, *Thysia ansorgii*, *Tilapia* spp.	High-quality flaked/pelleted foods; proprietary conditioning foods with higher concentration of vegetable matter; frozen bloodworms; glassworms; zooplankton; seasonally available livefoods, e.g. adult brineshrimp (*Artemia*) and water fleas (*Daphnia*); mosquito larvae. Two feeds daily, or more frequent smaller offerings.
Micropredators *Aulonacara* spp., *Chilotilapia rhoadesi*, *Chromidotilapia batesi*, *Cyathopharynx furcifer*, *Cynotilapia* spp., *Cyprichromis* spp., most small to medium-sized Malawi *Haplochromis* spp., the small *Hemichromis* spp., *Iodotropheus sprengerae*, *Julidochromis* spp., *Labidochromis* spp., small to medium-sized *Lamprologus* and *Telmatochromis* spp., *Nanochromis* spp., *Pelvicachromis* spp., *Tanganicodus, Spathodus* and *Eretmodus* spp., *Steatocranus gibbiceps, Teleogramma* spp., *Xenotilapia* and *Callochromis* spp.	High-protein flakes or pellets; freeze-dried or frozen zooplankton; frozen bloodworms; live *Daphnia, Artemia,* glassworms, mosquito larvae and snails. Three light feeds daily. Live foods encourage breeding.
Piscivores *Boulengerochromis microlepis*, *Cyphotilapia frontosa*, many large Malawian and Victorian *Haplochromis* spp., the large *Hemichromis* and *Lamprologus* spp.	High-protein pelleted foods; frozen or freeze-dried zooplankton; frozen whole smelt; thin strips of frozen fish fillets; live earthworms; small goldfish or guppies (useful for bringing fish into breeding condition); snails; bloodworms; adult *Artemia, Daphnia,* mosquito larvae. Two or three small daily feeds or one large daily meal.
Herbivores *Cyathochromis* spp., *Gephyrochromis* spp., *Labeotropheus* spp., *Oreochromis* spp., *Petrochromis* spp., *Petrotilapia* spp., *Pseudotropheus* spp., *Sarotherodon* spp., *Simochromis* spp., most *Steatocranus* spp., *Tropheus* spp.	High-roughage, low-protein foodstuffs; prepared flaked foods enriched with vegetable matter; algae; fresh vegetable foods, e.g. blanched baby marrows (zucchini), cooked green peas; weekly feed of frozen bloodworms or glassworms for animal protein. Several light feeds daily or two larger daily meals.

Routine maintenance

In an African cichlid tank, routine maintenance includes cleaning the front glass, maintaining the filter, removing waste particles from the substrate and making partial water changes. The frequency with which these tasks are carried out depends primarily upon a tank's population density, although algal growth is also affected by the amount of light the aquarium receives (see page 31).

Every three of four days, clean the inside surface of the front panel, using a magnetized scrubbing pad. Many algae species are easy to scrub away when newly settled onto a solid surface, but become progressively less easy to remove once established. Stubborn algae requires the attentions of a razor blade, a much messier and less agreeable process! A good growth of green algae on the back and sides of a tank plays a useful role in nitrogen cycle management and provides herbivorous cichlids with a valuable dietary supplement, so do not be too quick to remove it. Uncontrolled growth of blue-green algae results when a tank is overstocked and receives too much light. The only way of controlling it is by tackling the root causes of the problem.

Below: *A magnetized scrubbing pad is a convenient and effective means of removing algae growth from the front glass of the tank.*

Above: *Lake Malawi cichlids are extremely sensitive to nitrogen cycle mismanagement. Efficient biological filters, regularly cleaned and serviced, are essential if they are to thrive in captivity.*

Check filters daily. If you observe a reduction in the return flow, or the filter medium appears caked with waste, they are clearly in need of cleaning. Replace disposable media regularly and rinse reusable ones thoroughly under a stream of tapwater at the same temperature as that of the aquarium. There is no evidence that disinfectants, such as chlorine or chloramine, have any effect on their associated nitrifying bacteria. However, if there is a difference of more than one degree of pH and/or two degrees of hardness between tap and tank water, it is a good idea to rinse reusable filter media in water drawn from the aquarium. Abrupt changes in pH or hardness will damage the vital nitrifying bacteria.

Rinse the filter box clean and scrub out the intake and return tubes with a bottle brush. Organic deposits tend to accumulate inside the tubing over a period of time, reducing its diameter and thus the flow of water through the filter. If the unit is driven by a rotary magnetic impeller, remove the impeller and thoroughly scrub the interior of the well. An accumulation of organic residues

on the walls can seriously reduce filter efficiency.

African cichlids differ considerably in their tolerance of water changes (see page 27) and it is important to bear this in mind when planning and carrying out such a programme. Do remember that the more heavily stocked an aquarium, the more frequently it will require water changes, regardless of the species it houses. It is a simple matter to combine a water change with cleaning the tank bottom, as described on page 29. Unless a tank houses large, very messy cichlids, it is not necessary to vacuum the entire bottom with each water change. As long as the substrate is turned over once a month, serious problems are unlikely to develop.

The following checklist will simplify the task of monitoring an African cichlid tank:

● Are all the filters and airstones functioning normally? If not, why?

● Is the water temperature within the acceptable range for the tank's inhabitants?

● Is the tank's water clear, with a fresh smell?

● Is the number of fish visible equal to the number supposedly present in the tank? If not, why?

● Do the fish appear different in any way? Has their colour pattern changed? Is their rate of respiration elevated?

● Is the fishes' behaviour different in any way? Do they hang motionless below the water surface? Do they dash frantically about the tank at the slightest disturbance and try to hide? Do they refuse to eat when fed? Have the fish become more aggressive towards each other?

The answers to several of these questions are self-evident. Always restore filter or pump function immediately; adjust thermostat settings up or down; locate missing specimens; remove dead fishes from the aquarium and prevent obvious bullies from abusing their tankmates. Other symptoms require further investigation before the underlying problem – and appropriate solutions – can be identified.

Cloudy, greyish water, often associated with an unpleasant smell, indicates biological filter failure. It is often accompanied by a change in the appearance and behaviour of the fishes, such as an exaggerated rate of respiration, darker than normal coloration and 'panic' swimming in response to the slightest disturbance. 'Grey water' usually appears immediately after a tank has been set up. It is caused by adding a full

47

complement of fish to the aquarium before the filtration system is fully established. However, even a totally mature biological filter bed can 'crash' if its waste handling capacity is exceeded through overfeeding or by adding 'just one more' fish to the aquarium.

In the short term, you must address these symptoms immediately by removing any obvious decomposing organic matter – such as uneaten food or a dead fish – from the aquarium, making a 60-75 percent water change, thoroughly cleaning the substrate and cleaning the filter(s). It is also wise to add a chemically active medium to outside filters before refilling the tank with fresh water. In the long term there are two solutions to this problem: you can either reduce the waste load by cutting back on the number of fish stocked, or increase the filter's waste processing capacity by adding another unit to the system.

Aquarium water that remains clear, but assumes a yellowish or amber tinge may also indicate nitrogen cycle management problems. The first response to such a development should be to check the pH value of the water. An acidic reaction – especially in a sample taken from an aquarium where neutral to basic conditions should prevail – clearly indicates that something is amiss. Unless a new piece of driftwood has recently been added to the tank, the most likely cause of acidification is a build-up of metabolic waste. If a nitrite test yields readings of more than 0.01ppm and a nitrate test readings of more than 1ppm, you can assume that metabolite accumulation is the culprit.

Gradual acidification is most likely to occur in areas with soft

Below: *Regularly removing organic detritus from the substrate is simple using a siphonic gravel cleaning device such as this.*

Below: *Always treat 'raw' tapwater with a good-quality conditioner to reduce levels of chlorine and other 'disinfectants' in the water.*

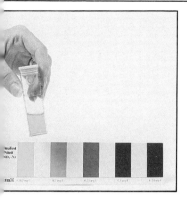

Above: *Reliable nitrite test kits enable the fishkeeper to monitor water conditions and implement corrective measures if necessary.*

tapwater. Apart from the possibility of nitrite poisoning, a drop in pH level clearly poses a more serious threat to the well-being of Rift Lake cichlids than to West African species. If you have chosen an alkaline substrate, such as coral gravel, or have added crushed coral or oyster shell to the filter and you make regular water changes, you should not be seriously troubled by it. More frequent water changes, combined with scrupulous attention to filter maintenance, will prevent these symptoms in the short run. In the long run, only reducing the number of fish housed in the tank or increasing filter capacity will resolve this problem.

Large-scale water changes tend to shift the pH into the alkaline range. Remember that such a shift converts harmless ammonium ions into toxic ammonia. To minimize the risk to the fish of ammonia poisoning, you can either add a chemically active medium to the filter system before adding fresh water to the aquarium, or use a one-step dechloraminating agent in conjunction with the water change (see also page 28).

Symptoms of acute respiratory distress, such as 'gulping' at the water surface, and marked colour pattern changes sometimes occur without an obvious cause, i.e. such as a pump or filter malfunction or loss of water clarity. In such a case it is worth investigating either biological filter failure or contamination of the water by some external agent, such as insecticides, household cleaning products or paint fumes. Before taking any corrective action, test the nitrite level. A positive test result (i.e. a concentration of more than 0.01ppm) indicates that excessive metabolite concentration is the probable cause of the problem and you can then implement the corrective measures outlined above.

A negative nitrite test, however, indicates that an external agent is at work. In this event, immediately remove and discard any activated carbon in the filter unit, add a chemically active medium to the system and carry out an 80-90 percent water change. If the fish can be saved, this should alleviate their obvious symptoms and give you sufficient time to locate and remove the source of the problem. Many insecticides are toxic in minute concentrations, so it may prove necessary to repeat the water change several times to eliminate all traces of them from the system. Once the symptoms have disappeared, remove the chemically active medium from the system and discard it.

The prognosis for victims of poisoning is not encouraging, hence the importance of preventing it. Never hang slow release insecticide strips or similar products in the same room as an aquarium. Never use insecticide aerosols near the aquarium, unless both the tank surface and its air supply are protected against accidental settling or uptake of the spray. The latter precaution applies equally to household cleaning agents and house paints. Finally, protect both dry driftwood and activated carbon from contact with toxic materials. They readily absorb such substances, only to release them into the water when added to an aquarium, with potentially lethal results.

Health care

Disease is a condition that entails the interaction of a suitable host (i.e. the fish), its environment (i.e. the aquarium) and a pathogenic (i.e. disease-causing) organism. Most fish pathogens are present at all times, but they cannot infect a potential host unless its normal defences have been weakened by environmental stress. Successful disease therapy, therefore, entails more than simply treating the obvious symptoms. If treatment is to prove successful, you must also improve the underlying environmental conditions that allowed the pathogen to invade its victim in the first place.

Abrupt changes in environmental conditions, including temperature and water chemistry, are well-known stress factors. One or the other is invariably implicated in outbreaks of 'ich' (caused by the protozoan *Ichthyophthirius multifiliis*), a parasitic disease. Poor nitrogen cycle management is of even greater significance. Long-term exposure to elevated concentrations of metabolic wastes severely weakens a fish's immune system, making it vulnerable to velvet and systemic bacterial infections, such as haemorrhagic septicemia and 'Malawi bloat'. Behavioural stress can have identical results. Removing sources of stress are, therefore, an integral part of any successful course of treatment. Simply adding medication to the aquarium in response to a particular set of symptoms will not solve the problem of disease in an African cichlid tank.

Correcting nitrogen cycle mismanagement may be simply a matter of adhering more rigorously to a schedule of partial water changes. More commonly, it means either reducing the number of fish in a given tank or upgrading its filtration system. Eliminating behavioural stress may also require thinning out a tank's population, but it may be enough to identify and remove an obvious bully, or simply to increase the amount of shelter available.

Parasitic infestations, including ich or flukes, respond favourably to a wide range of proprietary medications. Bacterial infections, however, are much more easily prevented than cured. The

Below: *The typical white spots of ich, or white spot, are clearly visible on this South American cichlid (*Geophagus hondae*). This condition responds well to proprietary treatments.*

Above: *These pitlike lesions on the head and along the lateral line are characteristic of 'hole-in-the-head' disease. Improve tank conditions and provide medicated foods for one week. (See table, page 52-3.)*

prognosis for cichlids so afflicted is guarded, even with antibiotic therapy. When this treatment is not available, recovery is highly unlikely. Under such circumstances, the most humane – and productive – course of action is to dispose of the afflicted fishes and concentrate on preventing the spread of the disease to those healthy individuals that remain.

There are several humane methods of destroying a terminally ill fish. The simplest is a swift slam against a solid surface, such as the

bottom of a bathtub. If you cannot contemplate such a direct approach, you can dispose of a small fish by leaving it in a cup of soda water at room temperature until there is no trace of gill movement. The dissolved carbon dioxide that causes the soda to effervesce acts as an anaesthetic overdose and simply puts the fish permanently to sleep. A tenfold dose of a commercially available fish tranquillizing agent will have the same effect. This approach is recommended for cichlids 10cm(4in) or larger.

Physical injuries
Treating physical injuries is part and parcel of African cichlid husbandry. These fish are all aggressive to a certain degree and,

Below: *A close-up of a single 'ich' parasite,* Ichthyophthirius multifiliis. *Up to 1mm(0.04in) across, with a characteristically curved nucleus.*

Below: *A microphoto of two gill flukes (*Dactylogyrus sp.*) clinging to a fish gill. Heavy infestations can cause respiratory distress.*

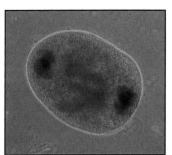

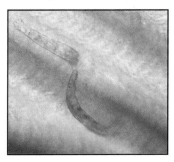

Common diseases of African Cichlids

Symptoms	Disease and causative pathogen
Discrete white spots on body and fins; scratching against solid objects.	'Ich' (White spot) caused by the ciliate protozoan *Ichthyophthirius multifiliis*.
Uniform dusting of minute gold dots on the body; respiratory distress. Most likely under soft, acidic water conditions.	Freshwater velvet caused by the single-celled dinoflagellate parasite *Oodinium pillularis*.
Repeated scratching against solid objects; thickening of mucus coat; respiratory distress. Most severe in overcrowded tanks.	Skin and/or gill flukes *Gyrodactylus* and *Dactylogyrus* spp.
Bloody streaks on the fin bases; open, bloody sores on the flanks.	Haemorrhagic septicaemia caused by various bacteria, including *Aeromonas*, *Pseudomonas* and *Haemophilus* spp.
Loss of appetite, with appearance of clear or whitish, mucus-like faeces; bloating of the abdominal cavity often follows.	'Malawi bloat', commonly the result of bacterial septicaemia (see above).
Erosion of the lateral line system of the head, resulting in the appearance of pitlike lesions.	'Hole in the head' disease. Irritative response to excessive levels of dissolved metabolities, often followed by secondary bacterial or protozoan infections.

Treatment

Remove environmental stress; treat with a commercial 'ich' medication according to the manufacturer's directions.

Improve nitrogen cycle management. Remove infected fish to a clean tank and treat with a commercial velvet medication according to the directions.

Reduce number of fish in affected tanks; treat with a commercial fluke medication strictly according to directions. Do not use formalin-based medications to treat fry.

Treat affected fish with a water-soluble broad-spectrum antibiotic (Oxytetracycline hydrochloride, kanamycin sulphate) according to directions; feed all exposed fish for one week with medicated food.

Treat affected fish with minocycline hydrochloride – 250mg per 38 litres (8.4 Imp. gallons/10 US gallons) of water; repeat after 2 days if initial treatment does not lead to resumption of feeding. Feed infected fish with proprietary medicated food as above. Removal of environmental stress is essential for successful treatment of systemic bacterial infections.

Reduce metabolite levels by lowering stocking rate and increasing frequency of water changes; treat secondary infections by feeding fish proprietary medicated food – both antibacterial and antiparasitic formulations – for a week. Erosion can be checked, but existing damage is irreversible.

in the confines of an aquarium, this tendency is apt to result in split or chewed fins and missing scales or similar skin damage. Fortunately, cichlids have remarkable recuperative powers and, as long as injured fish are given the correct supportive care, they can be expected to recover fully from truly appalling injuries.

Remove injured specimens to a hospital tank containing freshly drawn water of the same temperature and chemical make-up as their original aquarium. Treating superficial injuries, including torn or nipped fins, a few missing scales or small abrasions on the body or head, requires nothing more than the addition of water-soluble vitamin B_{12} to the hospital tank and raising the water temperature by a few degrees to promote healthy tissue regeneration.

Serious injuries, such as fins chewed back to their bases, open wounds or massive scale loss, require more careful attention. Several excellent proprietary remedies are available; be sure to use them strictly in accordance with the manufacturer's instructions. To prevent secondary bacterial infections, dose the hospital tank with a furan-ring based antibiotic and with a 'liquid bandage', as well as with soluble Vitamin B_{12}. This therapeutic regime offers the most hope that tissue regeneration will follow normally and result in minimal scarring or fin ray deformation.

As with accidental poisoning, serious injuries are easier to prevent than to treat. Serious fighting between cichlids is invariably the result of insufficient living space, a shortage of shelter, or a breakdown in the normal reproductive sequence. If you familiarize yourself with the space requirements of a given species before adding it to the tank and follow the suggestions outlined on breeding cichlids (see page 54 to 63), you should not, as a rule, find aggression an unmanageable or insoluble problem.

Breeding and rearing

From the outset we have sought to stress the great diversity of African cichlids and how important it is that the fishkeeper responds to their different needs in terms of housing and feeding. Their methods of reproduction are no less varied and some are easier to spawn in captivity than others. From the table on page 19 you can determine the mating system employed by each major group of cichlids, be it monogamy, harem polygyny, or open polygamy. You can then determine your approach to breeding them in captivity. Below, we examine each system in more detail and consider the management options that promote breeding success. Additional information on particular species is included in Part Two of this book.

Monogamous African cichlids

Most representatives of this group are substrate-spawners, but it also includes both primitive and advanced mouthbrooding species. What sets these cichlids apart from the two polygamous groups is the pair bond. This behavioural mechanism allows two aggressive – and often individually territorial – fishes to share the same space without incident for the duration of the breeding period. Success in breeding these cichlids depends entirely upon the formation and maintenance of a pair bond.

You cannot simply select a male and female at random, drop them together into their own tank and expect pairing to follow automatically. To begin with, both fishes need to be sexually mature, compatible and willing to breed. The presence of target fish to serve as an external focus for the aggressive behaviour of the prospective pair also plays an important role in sexual conditioning. Rearing to maturity a group of six to eight juvenile fishes of the same species is the most reliable way to obtain a compatible pair of monogamous cichlids. This approach works well precisely because it offers prospective pairs both a choice of partners and

plenty of suitable targets for their aggression. The same principle explains the ease with which a single male and female often pair up in a community setting. However, both approaches entail an element of risk to the other fish in the tank. Indeed, unless the aquarium is large enough to allow tankmates to escape the attentions of the newly established pair, it usually proves necessary to remove them for their own safety! These surplus fish can either be encouraged to pair off in turn, kept in a community situation, or sold to other aquarists or to retailers.

Target fish

The presence of 'enemies' also plays a role in maintaining the pair bond, although its importance varies greatly between species. The lamprologine cichlids of Lake Tanganyika have remarkably strong pair bonds that persist even in the absence of any target fish. The true tilapias, on the other hand, seem incapable of maintaining a pair bond in isolation. Most monogamous species fall somewhere between these two extremes.

The simplest, although not always the most practical, means of providing sufficient external stimuli to sustain the pair bond is to set up a pair in a tank large enough to afford other fishes a refuge beyond the limits of the breeding territory. This may entail doubling the minimum recommended bottom area of the breeding tank, so the feasibility of this option is in inverse proportion to the size of the cichlid in question! Fortunately, the mere sight of potential territorial rivals or spawn predators is enough to reinforce the pair bond. Placing the breeding tank next to a well-stocked aquarium, or isolating a target fish behind a clear partition in the breeding tank serves this purpose satisfactorily.

The breeding tank

Some cichlids are so aggressive that it is impractical to allow the two sexes unrestricted access to one another unless you can call upon the facilities of a public aquarium. In sexually dimorphic species, the easiest way to assure the female's safety is to separate the sexes with a barrier containing openings through which she can pass freely, but which restrict the movements of the larger male. Place the spawning site in the male's territory, which should be the larger area of the two. This 'separate compartment' approach gives the aquarist an opportunity to observe typical reproductive behaviour with minimum risk to the female, since it allows her to initiate contact with the male, as well as to break it off should he prove too rough a suitor. It also affords her a secure refuge should the pair bond break down after spawning has occurred.

The risk of death or injury to the female can be eliminated completely, albeit at the cost of the normal expression of the fishes' reproductive behaviour, by separating the sexes with a divider that permits free circulation of water between the two compartments of the breeding tank. This 'incomplete divider' approach exploits the fact that a male and female cichlid will perform their respective roles in

the spawning sequence without physical contact, as long as they are in sight of one another. Assuming there is a spawning site immediately adjacent to the barrier and good water circulation in the breeding tank, sufficient sperm will diffuse through the grid to fertilize a significant percentage of the female's eggs. Once the fry are free swimming, they can move freely between the two compartments, thus allowing each adult the opportunity to fulfill a parental role.

Although they differ over what constitutes a suitable spawning site, all these cichlids initially prefer a free choice of sites. Indeed, the selection process, no less than actual site preparation, appears to play an important role in reinforcing the pair bond in nature. Established pairs often develop a marked preference for a particular spawning site and when this happens, the breeder can adjust the tank furnishings accordingly.

A factor complicating the husbandry of some of the monogamous cichlids native to forest streams is that the sex ratio of their progeny is determined by the pH and hardness of their surroundings. This would not matter greatly if these fish, all representatives of the genera

Below: A 'separate compartment' tank is ideal for breeding isolated pairs of monogamous cichlids, such as Hemichromis guttatus. The openings in the divider allow the smaller female access to the male and a means of escape if he becomes too aggressive. Smooth stones offer a choice of spawning sites; floating plants provide cover.

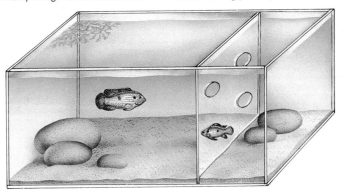

Pelvicachromis and *Nanochromis*, spawned only when pH and hardness values fell within an optimum range. Were this so, the breeder would always know precisely where he stood with regard to the outcome of the spawning effort. Unfortunately, these cichlids willingly spawn over a much broader range of pH and hardness values. The uninformed breeder is thus apt to find his endeavours crowned with unisex broods! Ignorance of this peculiarity of their natural history is probably the chief reason why so few *Pelvicachromis* species have become commercially established, notwithstanding the relative ease with which wild-caught fish can be induced to breed in captivity.

Premature respawning

Once the pair has spawned, the fishkeeper need usually do no more than provide food for the free-swimming fry. In the wild, effective custodial care places such demands on the time of the parent cichlids that most are unable to forage normally. They must thus subsist to a significant degree on body fat until their fry grow large enough to forage beyond the limits of the pair's original territory. This constraint on normal feeding precludes the possibility that the female will ripen another batch of eggs while tending fry. In captivity, thanks to the fishkeeper's generosity at feeding time, adult cichlids have no difficulty ripening a second batch of eggs while tending young from an earlier spawning.

For reasons that are not entirely clear, a few Tanganyikan species, notably *Lamprologus brichardi* and a number of *Julidochromis* species, are not restricted biologically from spawning again while still caring for their first brood. Not surprisingly, these cichlids have also evolved behavioural mechanisms that permit the adults to cope with several broods of fry within their territory. For most monogamous cichlids, however, such a situation

can lead to serious problems, as one or both adults tend to lose interest in the older fry with the onset of a new spawning cycle. The pair will often attempt to drive their first batch of fry out of the breeding territory as a prelude to respawning, regardless of the brood's ability to survive on it's own. If the young cannot move beyond the limits of the breeding territory, they are likely to be killed by their former guardians. In some instances, one parent – usually the female – will attempt to drive the older fry away, while the other parent attempts to defend them. This can lead to serious intersexual fighting and such premature respawning efforts doubtless account for many otherwise inexplicable breeding failures.

The simplest way to prevent such an eventuality is to reduce the temperature in the breeding tank to 22-23°C(72-74°F) as soon as the fry become mobile. This will slow their growth but, more importantly, it will also retard the maturation of a new clutch of eggs. Under such conditions, it is far more likely that the parents will tend their fry until they become fully independent. If you opt for higher tank temperatures and faster fry growth, be prepared to separate parents and offspring at the first signs of respawning, such as overt courtship behaviour or preparation of a spawning site.

Harem polygynists

The chief characteristic of fishes that adopt this mating system is that a single male monopolizes a group of two or more females, each of which defends a separate breeding territory against others of her sex. All the African representatives of this group are cave-spawners. Care of the eggs and yolk-sac fry is an exclusively maternal task, but male participation in the defence of the mobile young is not unusual and may be solicited by the female.

While there is a bond of sorts between a male and the members of his harem, it does not compare in intensity with the pair bond of monogamous cichlids. Females seem to accept any male as a spawning partner, while males will attempt to spawn with as many females as they can. Indeed, the more females present, the less danger his attentions pose to any one of them.

Breeding tank size is the most significant factor to influence reproductive success of haremically polygynous African cichlids in captivity. Obviously, it must be large enough to afford each female a territory of her own but, equally important, it must be large enough to afford a secure refuge for the male, whose proximity to the spawn is not appreciated during the initial phase of the reproductive cycle. Both the size of a female's territory and her willingness to tolerate the male's presence in the vicinity of the spawning site vary tremendously from one species to the next. Before setting up the breeding tank, you should carefully research the behaviour of a given species.

Apart from this consideration, you can handle these cichlids in the same manner as their monogamous relatives. The fry of many West African and Tanganyikan representatives of this group are acutely susceptible to nitrite poisoning. It is impossible to over-emphasize the importance of adhering rigorously to a programme of aquarium hygiene and making regular partial water changes in order to rear these fishes successfully.

Openly polygamous African cichlids
Representatives of this large assemblage of species lack the behavioural mechanisms that permit long-term cohabitation of the sexes. Contact between male and female is limited to the actual spawning act, and it is not unusual for an individual of either sex to

Below: *These open-spawning* Hemichromis guttatus *share the defence of their breeding territory and protect fry against predators.*

have multiple partners during the course of a single spawning effort. All openly polygamous African cichlids are maternal mouthbrooders, the male playing no role whatsoever in the care of the young.

Among the majority of openly polygamous species, the male defends a discrete breeding territory. The exceptions to this rule are the several *Tropheus* and *Cyprichromis* species of Lake Tanganyika and a number of Malawian *Haplochromis* of the open-water dwelling utaka group. Males of these non-territorial species actively attempt to separate a ripe female from other males, rather than lure her into a fixed territory. The persistence of territorial behaviour, no less than the size and spacing of breeding territories, varies considerably between representatives of this group in nature. These differences are less important in captivity, however, where space limitations restrict the expression of much species-typical behaviour. A single management approach, therefore, works equally well for all these cichlids under normal aquarium conditions.

Openly polygamous cichlids are best maintained either in groups of a single male with several females, or as single pairs in a community setting. Either approach results in an environment where the male is too preoccupied to harass a female to the point of killing her – the usual outcome if these cichlids are housed as isolated pairs. Serious breeders favour the first approach, as it maximizes fry production. However, most amateur cichlid enthusiasts would rather keep their fishes in a community setting, which allows them to maintain a larger number of species in the space available.

Aggression in these cichlids is intimately linked with reproduction. Thus, while it is unwise to try housing several males of the same species in an aquarium containing females, more than one male of the same species will live amiably

in a bachelor tank. Fish of the same species and sex are the most obvious targets of a territorial male's aggression, but it is not unusual for males of another species with a similar colour pattern to come in for unwanted attention as well. Consequently, it is not a good idea to house species with very similar male breeding dress together. A sexually active male grows progressively more belligerent towards tankmates, regardless of their identity, as his potential consort ripens her eggs. However, once spawning is over, the intensity of his territorial behaviour declines dramatically. In a community set-up, males will typically succeed one another as the tank's dominant resident, as their respective females attain reproductive readiness.

Once spawning has occurred, it is a good idea to remove the female to a nursery tank to complete the incubation sequence. As long as the main breeding tank contains sufficient cover to allow her to avoid immediate post-spawning harassment by the male, there need be no rush to isolate the female. With rare exceptions, maternal mouthbrooders are exemplary mothers, even under crowded conditions. However, isolation is essential if you wish to save any of the fry. While females will carry to full term in a community setting, they can seldom provide an effective defence of their mobile young in such an environment.

Provide a nursery tank in scale with the size of its intended occupant. A 20 litre(4.4 Imp. gallon/5.3 US gallon) tank will suffice for females up to 7.5cm(3in) long; a 38 litre(8.4 Imp. gallon/10 US gallon) tank for fishes measuring 10-15cm(4-6in); a 57 litre(12.5 Imp. gallon/15 US gallon) tank for fishes 15cm(6in) and larger. Fill it with freshly drawn water of the same chemical make-up and temperature as that of the spawning tank. Add a thin layer of gravel, as the light reflected from bare glass is very disturbing to these cichlids. The tank should contain shelter of some sort – a flowerpot or two facing away from the front glass will suffice – and a layer of floating plants. A mature sponge filter, a heater-thermostat and cover complete the furnishings.

Contrary to popular belief, many egg-carrying females will feed at reduced levels during the incubation period, if the opportunity presents itself. Supplementary feedings are not essential to their well-being and complicate nitrogen cycle management in the nursery tank. However, cessation of normal feeding does not mean that the normal metabolic processes also cease. Brooding females and their developing young generate significant quantities of nitrogenous wastes, so you should continue to make small but frequent water changes throughout the incubation period.

The fertilized eggs remain in the throat sac of the female for 10 to 28 days, depending upon the species in question and on the water temperature. At the end of this period, the female releases the fully developed young. In a few species, such as the *Labidochromis* of Lake Malawi, the fry then fend for themselves, but in most cases, the female continues to defend the free-swimming fry as they forage and will allow them to re-enter her mouth when seriously threatened.

Such prolongation of maternal care doubtless contributes significantly to fry survival in nature. However, in captivity it can actually threaten their well-being. Females often refuse to release their fry into what they perceive as unsafe surroundings, or they may be so nervous that even if they do release the fry they take them up again at the slightest disturbance. This overprotectiveness seriously interferes with the normal feeding behaviour of the fry. Thus most aquarists terminate the relationship between mother and young following their initial release.

Separating mother and fry is a simple matter in the case of females measuring 7.5cm(3in) or larger. Simply enfold the female in a net, hold her head-down over a container of water taken from the nursery tank, gently pull open her lower jaw and immerse her head in the water. The fry will immediately swim out of her open mouth. Females less than 7.5cm(3in) long are more difficult to manipulate safely. The following procedure sounds draconian, but is less stressful to both mother and young than more direct forms of

Separating a small female mouthbrooder from her fry

1 Insert female head down in the tube of a kitchen baster.

2 Insert lower third of baster into container of water from nursery tank. Remove baster from water.

3 Replace baster and squeeze bulb. Young are safely expelled into the water.

1

2   **3**

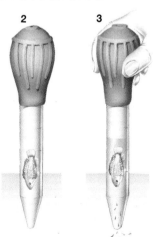

manipulation. Insert the female head-down in the tube of a kitchen baster. Replace the bulb and insert the lower third of the baster in a container of water from the nursery tank for a few moments. Remove the baster from the water for a minute or two, then replace it and give the bulb a few good squeezes. If no young are expelled from the baster's opening, remove it from the water for a few minutes and repeat the procedure. Respiratory distress will eventually force the female to make repeated gulping motions. The reverse flow of water produced by squeezing the bulb works with the female's jaw action to squirt the young harmlessly into the container.

Many breeders advocate a reconditioning period for the female before she is returned to the breeding tank. However, unless she is severely emaciated at the end of the incubation period, this is not essential. The fishkeeper can shake up the pecking order in the main tank before reintroducing the female by rearranging the tank furnishings and carrying out a large-scale water change. The female is thus spared serious harassment from her tankmates.

Artificial incubation of cichlid embryos

As a rule, cichlids are excellent parents. Nevertheless, it may transpire that a pair consistently devours its spawn, or that a female mouthbrooder fails to carry to full term. In most cases, the culprits are young, inexperienced fish and you need only exercise a bit of patience and the problem will correct itself. However, if the fish are extremely rare or valuable, you may opt for a less passive response to such behaviour. It is possible to incubate cichlid zygotes (fertilized eggs) artificially,

Below: *Young pairs of kribensis may eat their first spawns, but in time become excellent parents. Once the female brings the young out, both adults tend the fry.*

although there is a price to be paid in developmental casualties and congenitally deformed fry.

Artificially incubating the zygotes of the substrate spawning cichlids is a relatively straightforward matter. Transfer the leaf or rock on which the eggs were laid to a small aquarium filled with freshly drawn water of the same temperature and chemical make-up as that in the breeding tank. In order to duplicate the fanning behaviour of the parent female, place an airstone a short distance from the clutch and regulate the air flow to produce a moderate stream of bubbles. The bubbles must not come into contact with the spawn, nor should the current they produce be so strong that the developing young are battered by its flow. Failure to observe these conditions will result in a substantial percentage of congenitally deformed fry. The zygotes of cave-spawning species are light sensitive, so it is a good idea to cover the sides of their hatching tank with brown paper until the young become free swimming. A thermostatically controlled heater and secure lid are enough to complete the hatching tank's furnishings.

The developing young are vulnerable to bacterial attack. To minimize losses from this source, treat the water in the hatching tank with an antibacterial agent. Methylene blue is traditionally used for this purpose, but it oxidizes so rapidly that you must repeat the dosage several times while the eggs are developing if it is to retain its potency. Proprietary solutions of acriflavine hydrochloride, used according to the manufacturer's instructions, are a superior alternative to methylene blue. Satisfactory results can also be obtained with medications based on stabilized chlorine oxides.

Once all the fry have hatched, it is a good idea to change half the water in the tank. Repeat this procedure as soon as they become free swimming. Once the young have reached this point in their development, they should be treated in the same manner as fry left under parental supervision.

Artificial incubation of mouthbrooder zygotes is a more difficult proposition. A successful 'artificial mouth' must incorporate some system of gentle water circulation that mimics the tumbling the zygotes receive within the maternal throat sac. However, the current must not be so strong that the young are dashed against the sides of the container; such treatment will either kill them outright or cause serious congenital deformities. The easiest way to produce such conditions is to gently aerate the contents of a small container suspended in the heated water of a hatching tank to assure thermal stability.

It is essential to use some sort of antibacterial agent in the 'artificial mouth'. The preparations recommended in the artificial incubation of substrate-spawning cichlid zygotes are just as effective in the case of mouthbrooder fry. Bear in mind that in some instances, the dosages recommended by the manufacturer may prove harmful to mouthbrooder zygotes, so it is advisable to cut the recommended dose by half at the first attempt. If a higher concentration seems in order, increase the dose on subsequent occasions.

On alternate days, replace half the volume of water in the 'artificial mouth' with fresh water of the same temperature and chemical make-up. Remember to bring the concentration of antibacterial agent back up to strength following each water change. Check the progress of the zygotes frequently and remove any dead individuals as soon as you notice them. If they are allowed to decompose, further losses among the fry are inevitable.

However scrupulous your attention to hygiene, a certain percentage of mortalities is inevitable. The percentage depends primarily upon the state of development of the zygotes at the onset of artificial incubation.

The more time they have spent within the female's mouth the greater their chance of survival. Older zygotes are also less susceptible to congenital injury than unhatched eggs or youngsters at a very early stage of development. Once the young are fully mobile, they can be gently decanted into the hatching tank and offered food. Their rearing from this point on is a straightforward proposition.

Artificial incubation of mouthbrooder zygotes is a time consuming process and the yield of normal fry is invariably inferior to that resulting from normal maternal incubation. It is thus surprising that many breeders routinely 'strip' a female as early as three days after spawning to put her zygotes into an artificial mouth! On grounds of convenience alone, human intervention in cichlid brood care is best regarded as a last resort, rather than a routine matter.

Rearing African cichlid fry

Cichlid fry are easily reared. Those of *Anomalochromis thomasi* and some of the dwarf lamprologines require microworms for their first few meals. As soon as they are free swimming, the fry of other African cichlids can manage brineshrimp nauplii and finely powdered prepared foods. As improbable as it may seem, even adult fishes of robust species will snap up newly hatched brineshrimp with gusto. Always take this adult competition into consideration when calculating the quantity of brineshrimp nauplii to offer fry under parental supervision. Because of their small size, cichlid fry require a minimum of three feedings daily to prosper. Tanks that support a dense growth of green algae on their back and sides, or those that contain a fully matured sponge filter, offer the young a rich supplementary food source. It is not surprising that fry reared in such a setting grow more rapidly and evenly than those that rely exclusively on their keeper's offerings of food.

Above: *A striking collection of Tanganyikan cichlids that includes yellow* Lamprologus leleupi, L. tretocephalus, Tropheus duboisi *and* T. moorii. Lamprologus *fry grow relatively slowly;* Tropheus *fry grow more rapidly.*

Rigorous attention to aquarium hygiene is essential to the successful rearing of cichlid fry. They are extremely intolerant of dissolved metabolic wastes, yet the byproducts of their hearty appetites do nothing to simplify the task of sound tank management. Nitrogen cycle mismanagement can result in the loss of entire broods that succumb either to nitrite poisoning or to opportunistic diseases that attack the young when their immune systems have been compromised by chronic low-level exposure to

dissolved metabolites. The tank sizes suggested earlier for a single breeding pair or harem will usually afford adequate growing room for a brood of fry for four to six weeks after hatching. To sustain a satisfactory growth rate beyond this point, move them to larger quarters, or divide them between several rearing tanks.

By increasing the frequency of water changes, you can delay relocating the fry for a short time. However, not all species respond equally well to such a regime. In any event, as fry tend to be more sensitive to large-scale water changes than their parents, it is best to change a smaller volume of the water in their tank – between 10 and 25 percent every two or three days – than to make a 40-60 percent change every week. Young fry, in particular, are less

tolerant of dissolved chlorine than are larger specimens of the same species, so it is a good idea to use a commercial dechlorinating agent when making water changes in their tanks.

With the exception of the banded jewel fishes and the larger, piscivorous *Lamprologus* species, African cichlid fry are less given to sibling cannibalism than most of their New World cousins. The smaller brood sizes of most species also makes it easier to provide the fry with adequate growing room. One characteristic of the species bred to date is that male fishes grow more quickly. Bear this in mind when culling large spawns or choosing future breeding stock. If only the largest fry are selected for rearing to maturity, the likely outcome of the process will be a tank full of males!

Species section

Part Two features a representative selection of African cichlids. The fish are grouped by region of origin and introduced by a summary of the common features that influence their care in captivity. Details of the natural habitat, standard length, sexing, feeding type, compatibility and breeding behaviour of individual species follows. The first group features the riverine cichlids and includes those from Lake Victoria. Here are the jewel fishes, the *Chromidotilapia* lineage, tilapias and, of course, the haplochromines. They, together with the tilapias, are native to the rivers of eastern and southern Africa and dominate all the northern Great Lakes, other than Lake Tanganyika. In Lake Tanganyika, the second region to be considered, five distinct evolutionary lines of cichlids are represented. Here the lamprologines predominate, followed by the featherfins (*Ophthalmochromis*), with haplochromines a poor third. The largest and smallest

known cichlid species are native to Lake Tanganyika, and cichlids occur in every habitat within the lake and dominate many of them. Lake Malawi, the third region, is home to some 250 described cichlids, almost all haplochromines. It is considerably older than Lake Tanganyika and many of its cichlids have evolved a specialized life style, as exemplified by the rock-dwelling mbuna.

Research continues into the evolutionary relationships of African cichlids and recent revisions propose new generic names for cichlids included in the genera *Haplochromis* and *Lamprologus*. As these have not met with unanimous acceptance by interested researchers, we have chosen not to use *Cyrtocara* for the Malawian representatives of the genus *Haplochromis* and to treat the new names proposed for the extra-Malawian representatives as subgenera. Similarly, we have not incorporated revisions that would split *Lamprologus* into smaller genera.

Jewel fishes and their allies
The dwarf to medium-sized monogamous jewel fishes are unique among African cichlids in dividing long-term responsibility for their fry equally between the sexes. All do well in moderately hard (6-10° dH), neutral to slightly alkaline water and most thrive at temperatures of 21-27°C(70-80°F), increased to 30°C(86°F) for breeding. Although they tolerate short-term exposure to nitrite better than many West African cichlids, this is no excuse for sloppy nitrogen cycle management. Like other hardy cichlids, jewel fishes and their allies look better – and breed more readily – under a regime of effective biological filtration and frequent partial water changes. These cichlids may dig in conjunction with spawning, but do not pose a serious threat to well-rooted plants. They all appreciate a layer of floating plants. In a sufficiently large tank jewel fishes are easy to breed; at 27°C(80°F) eggs hatch after 40-72 hours and fry are mobile four days later. Offer them brineshrimp nauplii and finely powdered prepared foods. Remove fry from their parents four to six weeks after spawning, otherwise they may be eaten. When culling spawns or selecting fish for breeding, remember that males grow faster than females.

Anomalochromis thomasi

Dwarf jewel fish
● **Distribution:** Coastal rivers of Guinea, Sierra Leone and Liberia.
● **Habitat:** Clear and blackwater streams flowing under intact or recently disturbed rain forest.
● **Length:** Males up to 7.5cm(3in); females up to 5.7cm(2.25in).
● **Sexing:** Females often show a pattern of red dots in the shoulder region, as well as a pattern of bars on their flanks when courting. Males have longer soft dorsal and anal fins.
● **Feeding type:** Omnivore. Offer live food to encourage breeding.
● **Compatibility:** An excellent general community resident that poses no risk, even to small schooling fish. Pairs defend territories 60cm(24in) square, so it is possible to house several pairs in a well-planted tank 90-120cm(36-48in) long. May prove shy without floating plants.

Below: **Anomalochromis thomasi** *An ideal cichlid for beginners.*

Above: **Hemichromis elongatus** ♂ *A territorial male.*

Raising a group of youngsters to maturity together is the easiest way to secure a compatible pair. Spawns readily in a community tank, producing up to 500 eggs, but pairs seldom succeed in raising young to independence in the presence of fast-swimming schooling fish and scavengers. Prefers to spawn in a partially enclosed space and sometimes deposits its eggs on a large plant leaf. Hatching occurs in 40-48 hours at 27°C(80°F). The fry begin swimming about three days later, but do not absorb the remainder of their yolk-sacs for another 12-16 hours. Offer them microworms for the first three or four days, by which time they will have grown sufficiently to feed on brineshrimp nauplii and finely powdered prepared food. It is not unusual for young pairs to eat their first few spawns before settling down to model parenthood. The fry grow rather slowly, reaching 1cm(0.4in) after one month, even with good feeding and frequent partial water changes. They are more sensitive than other jewel fishes to nitrogen cycle mismanagement. Sibling cannibalism is not a problem in the dwarf jewel fish. Brood care persists for about four weeks after hatching and the young become sexually mature after eight to ten months, at 4.5cm(1.8in).

A slow-growing, but long-lived, hardy and easily bred dwarf cichlid. Aquarium strains are descended from the typical Sierra Leonian populations of *A. thomasi*. Liberian fish differ in details of their coloration and may represent a distinct species.

Hemichromis elongatus
Five-spot jewel fish

● **Distribution:** Coastal rivers in Guinea, Sierra Leone and southwestern Liberia; Togo southwards to northern Angola; lower Niger basin; Zaire (Congo) basin, including Shaba (Katanga) region; upper Zambezi basin.

● **Habitat:** Small streams, shallows of large rivers, marshes and oxbow lakes. Usually found close to submerged wood or aquatic plants. In the southern portion of its range, it penetrates coastal lagoons where the salinity can approach that of sea water.

● **Length:** Males up to 15cm(6in); females up to 10cm(4in).

● **Sexing:** Females are smaller and fuller bodied.

● **Feeding type:** Piscivore.

● **Compatibility:** Aggressive and violently intolerant of its own species. Devours any companion small enough to eat. Keep only one pair in tanks less that 180cm(72in) long. House with tilapias or large neotropical cichlids of a like disposition.

In a community setting the female is usually safe from injury by the male, but tankmates risk serious injury or death unless they can move out of the pair's territory. The pair bond tends to be unstable in the absence of target fish. The 'incomplete divider' method is suitable for breeding isolated pairs. In spawning sites close to cover they may produce up to 1000 eggs. The fast-growing fry are given to sibling cannibalism. They are sexually mature at about eight months at 7.5cm(3in).

Above: **Hemichromis guttatus** ♀ *The female is more intensely red.*

Hemichromis guttatus

Common jewel fish

● **Distribution:** Coastal rivers in Ivory Coast and Ghana west of the Volta; western Togo to southern Cameroons.

● **Habitat:** Small streams, marshes, oxbow lakes. Usually found close to submerged wood or aquatic plants.

● **Length:** Males up to 10cm(4in); females up to 7.5cm(3in).

● **Sexing:** Females are smaller, with redder base colour and less metallic blue spangling on body and fins.

● **Feeding type:** Micropredator. Offer colour food regularly.

● **Compatibility:** Relatively peaceful towards companions too large to eat, except when breeding. More tolerant of its own species than *H. elongatus*. Pairs defend territories 90cm(36in) square.

Pairs readily in a large community tank, with little risk of injury to tankmates. Rearing a group of young to maturity is the easiest way of securing a compatible pair. Pair stability is reinforced by the presence of target fish. The size difference between the sexes makes the 'separate compartment' method practical for breeding isolated pairs. The fish prefer to spawn in the open on smooth rocks or similar sites. Spawns can number 500 eggs. The fry grow rapidly and are less given to sibling cannibalism than those of the five-spot jewel fish. They attain sexual maturity six to eight months after hatching, at 5cm(2in) for males, 3.75cm(1.5in) for females.

Thysia ansorgii
Five-spot cichlid

● **Distribution:** Coastal rivers from just west of the Bandama in the Ivory Coast to the Ankombrah in southwestern Ghana, thence from the Ouémé River in Benin to the neighbourhood of Douala in the Cameroons.

● **Habitat:** Small streams flowing under intact or recently disturbed forest cover. Sometimes found in swamps and oxbows, usually in association with stands of aquatic plants.

● **Length:** Males up to 10cm(4in); females up to 7.5cm(3in).

● **Sexing:** Easily sexed. Males have long, flowing soft dorsal and anal fins, as well as an asymmetrical caudal fin with a distinct reticulate pattern. Females have a small cluster of iridescent white scales just above the vent.

● **Feeding type:** Omnivore.

● **Compatibility:** A 'giant' dwarf cichlid whose behaviour in captivity resembles that of *A. thomasi* in all significant respects. Introduce appropriate dither fish to overcome shyness. See page 38 for advice on suitable tankmates.

Breeding details as recommended for the dwarf jewel fish (see page 66), but pairs are usually more successful in raising fry in a community setting than *A. thomasi*. Prefers to spawn in a partially enclosed space and will often deposit its eggs on a vertical surface. Spawns can number up to 250 stalked eggs, which hatch in 72 hours at 27°C(80°F) and the fry are mobile seven to eight days later. The fry are more sensitive to dissolved metabolic waste than are those of the red jewel fishes, but are not difficult to rear under a routine of frequent partial water changes. The young are sexually mature six to seven months after hatching, at 6.5cm(2.5in) for males and 5cm(2in) for female fishes.

Below: **Thysia ansorgii** ♀ *Note the iridescent white ventral spots.*

The *Chromidotilapia* lineage

This large group of West African cichlids consists of the genera *Chromidotilapia*, *Pelvicachromis* and *Nanochromis*. These are dwarf to medium-sized cichlids with clear visual distinctions between the sexes. Males are larger and have longer, more elaborate finnage. Females are smaller and more colourful. A brilliantly coloured ventral spot and contrasting, often metallic, coloration in the spiny dorsal fin are typical of the female's courting dress. In all these species, the female's courtship display centres around the presentation of this vividly coloured ventral blotch to the male.

All representatives of this group are extremely sensitive to dissolved metabolites. Scrupulous attention to proper nitrogen cycle management is absolutely essential to their successful maintenance. Rapids-dwellers, such as *Nanochromis*, require high levels of dissolved oxygen to prosper and appreciate strong water movement. Ideally, therefore, combine an outside power filter with brisk aeration in their tanks. Many *Pelvicachromis* and *Nanochromis* species hail from soft, acid-water habitats, but in captivity most do well – and will even spawn – in slightly hard (6-10°dH), neutral to slightly alkaline water. However, their progeny will display a balanced sex ratio only if the pH and hardness levels in the spawning tank fall within a narrow range. A temperature range of 21-27°C(70-80°F) suffices for day-to-day management, with an increase to 30°C(86°F) to encourage breeding.

These cichlids do not eat aquatic plants, but the normal foraging behaviour of *Chromidotilapia* and the larger *Nanochromis* species poses some risk even to well-rooted specimens. Potted plants are usually immune from such disturbance. Digging is restricted to periods of sexual activity and limited to the vicinity of the spawning site. All appreciate the security of a screen of floating plants in the aquarium, such as *Ceratopteris cornuta*.

Chromidotilapia guntheri
Mouthbrooding krib
● **Distribution:** Coastal rivers from Ivory Coast to the southern Cameroons, including the Niger and Volta basins.
● **Habitat:** The slower stretches of streams and rivers, usually over relatively fine substrates and often close to stands of submerged wood. Less common in oxbows or lakes.
● **Length:** Males up to 18cm(7in); females up to 15cm(6in).
● **Sexing:** Females are smaller, with a rosy violet ventral blotch and a nacreous gold spiny dorsal fin.
● **Feeding type:** Omnivore.
● **Compatibility:** Will eat fish as large as a female guppy, but overall, an inept predator. Sexually inactive individuals ignore larger midwater-swimming species. Pairs defend territories 100cm(39in) square. They will disregard tankmakes that move beyond their territorial borders.

Rearing a group of young to maturity together should produce a compatible pair. Pair stability is reinforced by the presence of target fish. The size difference between the sexes makes the 'separate compartment' breeding approach workable with isolated pairs. *C. guntheri* is an advanced mouthbrooder that deposits its eggs on a flat surface and fertilizes them in conventional cichlid fashion. The male then takes them into his mouth. Some pairs share the incubation sequence, while in others the male is the sole custodian of the brood until, at 27°C(80°F), the mobile fry emerge, a week after spawning. Both

Above: **Chromidotilapia guntheri** ♀ *Rosy area reddens at spawning.*

parents then share custodial duties, to the point of allowing the young to take shelter in their mouths should danger threaten. Broods rarely exceed 100 fry. The young can take brineshrimp nauplii and finely powdered prepared food for their initial meal. The fry grow rapidly and can measure nearly 2.5cm(1in) within a month of their initial release. This is the time to separate them from the adults, since they have grown too large to fit within their parents' mouths. The young attain sexual maturity at eight to ten months after hatching, at 10cm(4in) for males, 6.5cm(2.5in) for females.

This hardy, medium-sized cichlid prospers over a wide range of pH and hardness conditions and makes an excellent addition to a community of like-sized African cichlids. The mouthbrooding krib is known in the older aquarium literature as *Pelmatochromis guentheri*.

Nanochromis parilius
Nudiceps
● **Distribution:** The lower Zaire River, from the Malembo (Stanley) Pool to its mouth.
● **Habitat:** Rippling pools of small streams flowing into the main channel of the Zaire River, as well

as calm stretches of the main river itself, typically over coarse gravel.
● **Length:** Males up to 7.5cm(3in); females up to 5.7cm(2.25in).
● **Sexing:** Females are much deeper bodied, with shorter soft dorsal and anal fins.
● **Feeding type:** Omnivore. Live foods encourage breeding.
● **Compatibility:** Males are very intolerant of other males of their own species. Even sexually inactive individuals of both sexes often behave aggressively towards other small bottom-dwelling fish, but will ignore midwater-swimming tankmates too large to make a comfortable mouthful. Best housed in tanks at least 90cm(36in) long, as males are often very aggressive towards sexually unreceptive females.

Even sexually quiescent females appear ripe, for a conspicuous ovipositor is visible at all times. However, truly ripe individuals appear ready to burst. This species is best maintained on a harem basis to reduce the risk of injury or death to the female during courtship. If only a single pair is available, it is essential to provide an abundance of hiding places and suitable target fish in the breeding tank. Courtship is protracted and it is not uncommon for a young

Above: **Nanochromis parilius** ♀ *Has a rounder ventral profile.*

female to eat her first few spawns. As with the common krib, the initial phases of brood care are exclusively maternal and the female's refusal to allow the male to enter the breeding cave is a reliable indication that spawning has occurred. Spawns can number up to 100 eggs, but are usually smaller. Hatching occurs in about 48 hours at 27°C(80°F) and the female brings the free-swimming fry out of the cave five days later. At this point, the male may join his mate in fry care. The robust young can take brineshrimp nauplii and finely powdered prepared food for their initial meal. The fry are very sensitive to high ambient bacterial and dissolved metabolic waste levels. Twice-weekly partial water changes of 50 percent during the first month help to avoid losses and maximize growth. Parental care lasts about four weeks in captivity. The fry are sexually mature between six and eight months after spawning, at 6cm(2.4in) for males, 4cm(1.6in) for female fishes.

Pelvicachromis pulcher
Kribensis; Common krib
● **Distribution:** Coastal rivers of southern Nigeria, from the Niger delta westward to the Benin border.

● **Habitat:** Small streams and the slower stretches of rivers, usually over relatively fine substrates and always closely associated with stands of aquatic plants.
● **Length:** Males up to 10cm(4in); females up to 7.5cm(3in).
● **Sexing:** Females are smaller and deeper bodied, with a bright red ventral blotch and a metallic gold spiny dorsal fin.
● **Feeding type:** Micropredator. Live foods encourage breeding.
● **Compatibility:** An excellent general community resident, but may prove shy unless kept in a well-planted tank with small schooling fish. Pairs defend territories 60cm(24in) square, so it is possible to house several pairs in a tank 90-120cm(36-48in long). Restricts digging to spawning site.

A single male and female will pair readily in a community setting, but rearing a group of young to maturity remains the surest route to a compatible pair. Pair stability and parental reliability are reinforced by the presence of target fish. Courtship is protracted and pairs often go through several false spawns before finally producing a clutch of eggs. Young pairs may eat their first few spawns, but eventually settle down to model parenthood. Up to 100 eggs are placed on the roof and sides of an enclosed site and tended exclusively by the female. Her refusal to allow the male to enter the spawning site is a clear indication that spawning has occurred. The eggs hatch in 48-50 hours at 27°C(80°F) and the female brings the free-swimming fry out of the cave five days later. At this point, both parents diligently care for their progeny. The young can take brineshrimp nauplii and finely powdered prepared food for their first meal. Rearing them poses no problems if they are kept under a routine of frequent partial water changes. Parental care lasts from four to six weeks in captivity. A pH of 6.8-7.2 will result in a balanced sex ratio among the young, which can be reliably sexed on the basis of differences in ventral fin shape four to five months after spawning. They attain sexual maturity three months later, at 6.5cm(2.5in) for males, 4.5cm(1.8in) for females.

Pelvicachromis sacrimontis apparently ranges from the Niger Delta to the Cross River, replacing *Pv. pulcher* in eastern Nigeria. It is imported and marketed under the name of giant krib, and is immediately recognizable by the absence of metallic gold coloration in the spiny dorsal fin of either sex. Maintenance requirements as for *Pv. pulcher*, but a pH of 5.5-6.0 is required to produce a balanced sex ratio among the fry.

Below: **Pelvicachromis pulcher** *Fin markings vary in males.*

Tilapias and their allies

Representatives of this economically important cichlid lineage occur throughout Africa and even into the Near East. A few *Oreochromis* species are endemic to one or other of the Great Lakes, and Barombi Mbo, a crater lake in the Cameroons, supports a unique collection of tilapia-descended genera. However, the majority are river dwellers, so it seems appropriate to consider them under that heading. Most tilapias are medium-sized to quite large cichlids, although a few dwarf species are known. As juveniles, they all show the typical black 'eye-spot' or 'tilapia-spot' at the base of the soft dorsal fin. This spot disappears from the colour pattern of mouthbrooding representatives of the group, but usually persists into adulthood among the substrate spawners.

Despite their hardiness and attractive coloration, tilapias do not enjoy a great deal of popularity among aquarists. Most tilapias are omnivores with strong herbivorous tendencies, and their habit of treating planted aquariums as self-service salad bars is not appreciated by most aquarists.

Steatocranus and *Chaytoria* species are sensitive to dissolved metabolites and require the same attention to proper nitrogen cycle management as representatives of the *Chromidotilapia* lineage. The remaining tilapias can tolerate elevated nitrite levels for considerable periods of time without suffering ill-effects although, like all hardy fishes, they look their best – and are more likely to spawn regularly – under a regime of frequent partial water changes. Tilapias as a group are indifferent to the chemical make-up of their water as long as extremes of hardness or pH are avoided. Most *Tilapia*, *Sarotherodon* and *Oreochromis* species appreciate a temperature range of 21-27°C(70-80°F) for day-to-day maintenance, with an increase to 30°C(86°F) to encourage breeding.

Oreochromis mossambicus

Mozambique mouthbrooder
● **Distribution:** Coastal basins from Quelimane in southern Mozambique southward to the Pongola River in South Africa. South of this point, this species occurs in brackish water habitats as far as Algoa Bay.
● **Habitat:** The lower reaches of rivers in areas of moderate current, as well as in coastal lagoons, oxbow lakes and swamps. Typically found over soft bottoms, often in well-planted habitats.
● **Length:** Males up to 38cm(15in); females up to 25cm(10in).
● **Sexing:** Females are smaller than males and lack their bright colours and excavated cranial profiles.
● **Feeding type:** Omnivore with strong herbivorous tendency.
● **Compatibility:** Males defend a territory centred on a huge spawning pit with a diameter roughly twice their own overall length. Adult males are aggressive towards males of the same species and other cichlids with a similar colour pattern. Single individuals make good cichlid community residents if kept with tankmates of a like size and disposition. Although it will eat fish small enough to be easily swallowed, *O. mossambicus* is a clumsy predator.

Short of physically separating the sexes, there is no way to prevent this species from breeding in captivity. Males attempt to lure ripe females into their spawning pit with a vigorous 'headstand' display. The two fish then begin a reciprocal circling pattern that leads to the expression of a batch of eggs by the female. She immediately picks them up and nibbles at the male's vent to elicit

Above: **Oreochromis mossambicus** ♂ *Male digging a nest pit.*

ejaculation. The eggs are thus fertilized in her mouth. Once the female is spent, the male chases her from his territory and attempts to repeat the performance with another female. The incubation period lasts 14 days at 27°C(80°F) and, in nature, the female will continue to protect the mobile fry for an additional week. A large female can produce up to 1700 eggs, but spawns of 150 to 200 fry are the usual rule in captivity. The robust fry can take fine prepared foods and brineshrimp nauplii for their first meal. They are easily reared, with the males growing faster and bigger than females and both sexes reaching sexual maturity relatively easily. With heavy feeding and frequent water changes, it is not unusual for the young to begin breeding when only six months old, at a length of 5.75cm(2.25in) for males, 4.5cm(1.8in) for females.

Steatocranus casuarius

Bumphead/Buffalohead cichlid
● **Distribution:** The lower Zaire River, from the Malembo (Stanley) Pool to Matadi.
● **Habitat:** Stretches of relatively calm water between rapids in the main channel of the Zaire River, typically over rocks.
● **Length:** Males up to 16.5cm(6.5in); females up to 10cm(4in).
● **Sexing:** Females are smaller and lack the male's prominent

nuchal hump and long, flowing soft dorsal and anal fins.
● **Feeding type:** Herbivore.
● **Compatibility:** Somewhat aggressive towards its own species and other bottom-dwelling cichlids. It is possible to keep more than a single pair in a large aquarium 120cm(48in) long, generously furnished with caves and similar hiding places. They are apt to be shy unless housed with suitable target/dither fish, which are ignored outside periods of reproductive activity. These cichlids move large volumes of gravel when preparing a spawning site, but will not usually disturb potted plants.

A monogamous, cave-spawning cichlid that pairs and spawns readily in a community setting. Courtship is not obvious. A sudden increase in digging activity followed by the appearance of a short white ovipositor about 24 hours before spawning are the most reliable indications that reproduction is in the offing. Up to 100 large, pear-shaped eggs are placed on the roof and sides of a cave. The division of parental roles is comparable to that in the common krib (see page 72). The eggs hatch after five days at 27°C(80°F), and the robust fry emerge from the cave four to five days later. Treat them as recommended for the fry of *N. parilius* (see page 72). The adults are extremely protective of their fry and usually have no difficulty rearing them to independence in a community tank, although this approach to breeding can be rather hard on its other residents. Parental care persists until the female ripens a new clutch of eggs, between six and eight weeks after spawning. If the breeding tank is large enough, there is no need to remove the fry at this point. They do not prey on their younger siblings and, in consequence, the fry are tolerated within the breeding territory by their parents. The young are sexually mature about six months after spawning, at 5cm(2in) for males slightly less for females.

Above: **Steatocranus casuarius**
♂ *Has prominent nuchal hump.*

Below: ♂ *Only the male sports long soft dorsal and anal fins.*

Above: **Tilapia mariae** ♀ *Clearly showing five spots.*

Tilapia mariae
Five-spot tilapia
● **Distribution:** From the Bandama River in Ivory Coast to the Bia River in southwestern Ghana; thence from Lake Toho in Benin to the Kribi River in the southern Cameroons.
● **Habitat:** The lower reaches of rivers flowing under forest cover, as well as in coastal swamps and lagoons. Abundant in deep, flowing pools where submerged wood, rocks and stands of algae-coated aquatic plants are plentiful.
● **Length:** Males up to 30cm(12in); females slightly smaller.
● **Sexing:** Not easily sexed. Males grow slightly larger than females, while the red shoulder spot of the female is usually brighter. Large females often have a red 'bib' on the chest immediately in front of the ventral fins. Males are more aggressive towards their companions, whereas females tend to behave submissively.
● **Feeding type:** Omnivore with strong herbivorous tendencies.
● **Compatibility:** Adults are extremely aggressive towards their own species. Single individuals make good cichlid community residents if kept with tankmates of a like size and disposition. House only one pair per tank. They may make life extremely unpleasant for their companions, regardless of

species, with the onset of reproductive behaviour.

Handle this monogamous species in the same manner as the common jewel fish (see page 68). Breeding efforts are most likely to succeed with young pairs in a tank 120cm(48in) long. Spawns numbering as many as 5000 eggs are often placed within a partially enclosed space or on a vertical surface. The female is the primary caretaker of the developing zygotes, while the male defends the pair's territory against intruders. The eggs hatch in 60-72 hours at 27°C(80°F). The female may move the fry several times to pre-dug pits before they become mobile three days later. Both parents vigorously defend the free-swimming young. The fry are easily raised on a diet of brineshrimp nauplii and powdered prepared foods. With frequent water changes, they reach 4cm(1.6in) after 12 weeks. Parental care can persist at least this long in captivity, but most aquarists prefer to separate the fry from their parents by the end of the second month after spawning. Like the young of most tilapias, those of *T. mariae* are sexually precocious. Reproductive activity begins between six and eight months after spawning, at 6.5cm(2.6in).

Haplochromis and their allies

The great majority of these cichlids are small to medium-sized fish. All are openly polygamous, maternal mouthbrooders, in which the males grow faster and bigger than females and both sexes reach sexual maturity relatively early. In most species the sexes are easily distinguished; males are brilliantly coloured and females are drabber. In the great majority of *Haplochromis* species, the male sports conspicuous pale yellow to orange spots on the anal fin about the same size as that species' eggs. During spawning, the male displays these pseudo-ocelli conspicuously to the female, who attempts to take them into her mouth, just as she would her unfertilized eggs. This mouthing behaviour triggers ejaculation, so that in trying to pick up the 'egg dummies' she collects a quantity of sperm that fertilizes the eggs already in her mouth. *Pseudocrenilabrus* and some species native to the rapids of the Zaire basin lack such pseudo-ocelli. Their spawning behaviour resembles that of *Oreochromis mossambicus*, with the egglaying female nibbling directly at the male's vent.

These cichlids are essentially indifferent to water chemistry, but dislike extremes of pH and hardness. Specialized rapids-dwelling species and those native to the northern Great Lakes are sensitive to dissolved metabolites and require regular partial water changes. Haplochromines from quieter river waters and isolated shore pools of large lakes can tolerate elevated nitrite levels for considerable periods, although such hardiness is no excuse for inadequate nitrogen cycle management. Most extra-Malawian haplochromines tolerate temperatures as high as 35°C (95°F) if their tank is well aerated. Although the incubation period is shorter in warmer water, males are also considerably more aggressive at higher temperatures. For daily maintenance, temperatures of 21-24°C(70-75°F) will suffice, with an increase to 27-30°C(80-86°F) to encourage breeding.

It is unfortunate that these haplochromines have been somewhat overshadowed by their distant Malawian cousins. They are just as colourful, will thrive and breed in smaller tanks and all are indifferent to rooted aquatic plants. Not always available, but worth taking some trouble to find.

Haplochromis (Astatotilapia) burtoni

Burton's mouthbrooder
● **Distribution:** The basin of Lake Tanganyika, including the Malagarasi River and associated swamps.
● **Habitat:** Small streams, swamps and isolated shore pools along the coast of Lake Tanganyika. Uncommon in the lake itself, which seems to serve chiefly as a dispersal corridor to preferred peripheral habitats.
● **Length:** Males up to 10cm(4in) and females up to 7.5cm(3in) in captivity; 6.5cm(2.5in) and 5cm(2in) in nature.
● **Sexing:** Males are brilliantly coloured; females are drabber.

● **Feeding type:** Omnivore. Offer colour food regularly.
● **Compatibility:** Males are very aggressive towards one another and towards other male haplochromines with a similar colour pattern. Very large males are too bellicose to be kept together in tanks less than 150cm(60in) long. Groups of several younger males can be maintained successfully in tanks more than 120cm(48in) long. Midwater-swimming fishes too large to make a convenient mouthful are ignored. Since males move extensive areas of gravel during the construction of a nest pit, potted plants are less likely to be disrupted than rooted ones.

Above: **Haplochromis burtoni** ♂ *Male has conspicuous 'eggspots'.*

H. burtoni is one of the easiest African cichlids to breed. Spawns can number up to 120 eggs, but 50 is closer to the norm. The incubation period lasts 10 days at 30°C(86°F). The fry are large enough to take brineshrimp nauplii and finely powdered prepared foods for their initial meal. They are easy to rear when handled in the same fashion as those of *Oreochromis mossambicus* (see page 74). This species matures remarkably quickly; with frequent feedings and water changes, the young can attain sexual maturity eight weeks after release, at a length of 3.75cm(1.5in) for males, 2.5cm(1in) for females.

Haplochromis (Prognathochromis) pellegrini

Green arrow haplochromis
● **Distribution:** Lake Victoria.
● **Habitat:** Vegetated inshore habitats throughout the lake.
● **Length:** Males up to 10cm(4in);

females slightly less.
● **Sexing:** Males are brilliantly coloured; females are drabber.
● **Feeding type:** Piscivore.
● **Compatibility:** Males behave aggressively towards one another. Several individuals will coexist in a tank more than 150cm(60in) long, but only one will develop the typical courting dress of the species and enjoy reproductive success. Females can also be somewhat snappy towards one another if crowded, and individuals of both sexes may react unfavourably to other haplochromines with the same overall appearance. Will prey efficiently upon fish as large as a female guppy, but ignores larger midwater-swimming companions. May excavate a nest pit, but not as energetically as *H. burtoni*.

Do not attempt to maintain a breeding group in a tank less than 120cm(48in) long. Apart from its need for more swimming room, *H. pellegrini* requires the same

Below: **Haplochromis pellegrini** ♀ *A ripe female.*

management as other small, openly polygamous cichlids. Spawns can number up to 75 eggs, but 35 is more usual. The incubation period lasts 14 days at 30°C(86°F). Offer the fry brineshrimp nauplii and finely powdered prepared foods for their first meal. Given scrupulous attention to cleanliness, they are easy to rear and can be reliably sexed on the basis of colour differences 12 weeks after release. However, reproductive activity does not commence until the young are at least six months old, at a length of 7cm(2.75in) for males, slightly less for females.

Pseudocrenilabrus multicolor

Egyptian mouthbrooder
● **Distribution:** The Nile River drainage from Lake Albert and the Murchison Falls northwards to the Delta.
● **Habitat:** Swamps, small streams and the peripheral waters of rivers and lakes, usually in heavily planted habitats.
● **Length:** Males 6.5cm(2.5in); females 5cm(2in).
● **Sexing:** Females are drabber.
● **Feeding type:** Omnivore. Provide live food and colour food.

● **Compatibility:** Males are very aggressive towards one another at all times and, with the onset of breeding, prove surprisingly belligerent towards other bottom-dwelling fish. Breeding territories measure about 45cm(18in) square, which are very large for such a small fish. Usually does poorly when housed with larger *Haplochromis* species, but does surprisingly well in the company of *Oreochromis* species. Ignores midwater-swimming companions too large to be easily swallowed.

The Egyptian mouthbrooder is very easily bred. As males court vigorously, this species is best housed in tanks more than 90cm(36in) long. Spawns can number 100 eggs, but 35-50 is closer to the norm. Incubation lasts 10 days at 30°C(86°F). Young females may fail to carry their initial brood to term, but can be counted upon to complete subsequent efforts successfully. Raise the fry on a diet of brineshrimp nauplii and powdered prepared foods. With frequent water changes, they can attain sexual maturity four months after spawning, at a length of 2.5cm(1in) for males slightly less for females.

Below: **Pseudocrenilabrus multicolor** ♂ ♀ *The male is shown above.*

LAKE TANGANYIKA CICHLIDS

Despite the wide diversity of size, form and reproductive patterns among the Lake Tanganyikan cichlids, they share sufficient features to warrant a common approach to their care in captivity. All require hard, alkaline water to prosper and all are extremely sensitive to abrupt changes in both the chemical composition and temperature of their water. Most species prefer temperatures in the range of 22-24°C(72-75°F) for day-to-day maintenance, with an increase to 27-30°C(80-86°C) for breeding. Without exception, these cichlids are highly susceptible both to ammonia and nitrite poisoning. They dislike large-scale water changes, so the fishkeeper must rely on light stocking rates, efficient biological filtration and small (10-15 percent), frequent water changes for effective nitrogen cycle management.

Lamprologus and their allies

Some 48 species of *Lamprologus* from Lake Tanganyika have been described and exported as ornamental fish, and all but the giant predator *Boulengerochromis microlepis* have been well received by aquarists. *Cyphotilapia* and the larger *Lamprologus* species are formidable piscivores, whereas *Telmatochromis* and the medium-sized *Lamprologus* are micropredators. *Chalinochromis* and *Julidochromis* employ their specialized, tweezer-like teeth to pluck invertebrates from the algal and sponge mats that encrust the lake's rocky substrate. Their diet in captivity should reflect this reliance on animal food. These cichlids dig when breeding, but restrict their excavations to the immediate vicinity of the spawning site and pose little threat to rooted plants, such as Java or African fern.

Chalinochromis, *Julidochromis* and most *Lamprologus* and *Telmatochromis* species are monogamous in nature. Most of the smaller ostracophil or shell-dwelling, *Lamprologus* and *Telmatochromis* species are harem polygynists. However, many species known to practise monogamy in the wild tend to shift to a polygynous system in captivity if a surplus of females is present. As a rule, pair (or harem) bonds are very strong in lamprologines, so it is possible to maintain breeding pairs or groups in isolation with minimal risk to the female. Courtship is not usually overt and pairs are very secretive about spawning. The female's reluctance to leave the cave usually indicates that spawning has occurred, but often the first indication is the appearance of free-swimming fry! Eggs hatch in 40 to 48 hours at 27°C(80°F); the fry are mobile five days later.

Save for *Cyphotilapia*, a maternal mouthbrooder, these substrate-spawning cichlids prefer enclosed spawning sites. Both parents share responsibility for fry, much like the *Pelvicachromis* species. However, the young are protected only while they remain within their parents' territory and no effort is made to retrieve any that wander further away. In other respects, lamprologine reproductive behaviour displays great sophistication. Several species practise group-spawning and communal defence of the resulting fry. Sometimes the fry from a previous brood remain within the parental territory to help protect younger siblings. At one time it was thought that only 'higher' vertebrates, such as birds or mammals, exhibited such 'advanced' behaviour.

All lamprologines are skilful jumpers and require tightly covered aquariums.

Above: **Julidochromis marlieri** ♂ *Striking body pattern.*

Julidochromis marlieri

● **Habitat:** A widely distributed, albeit uncommon, inhabitant of the rocky shore to depths of 35m(115ft).

● **Length:** Males up to 15cm(6in); females slightly less.

● **Sexing:** Males grow slightly larger than females and sport a conspicuous, penis-like genital papilla that inexperienced aquarists often mistake for an ovipositor.

● **Feeding type:** Micropredator.

● **Compatibility:** One pair defends a territory about 60cm(24in) square. It is possible to keep several pairs in the same aquarium as long as each can secure a territory. Individuals may behave aggressively towards the closely related *J. regani*, but otherwise prove good neighbours to Tanganyikan cichlids of their own size or slightly smaller. *Julidochromis marlieri* typically ignores midwater-swimming tankmates.

These monogamous cichlids pair readily, although the pair bond is not as strong as in most other lamprologines. Keep a few schooling fish with isolated couples to act as target fish and reinforce the pair bond. Like other *Julidochromis*, this species can either produce a large clutch of up to 300 olive-green eggs every four to six weeks, or spawn 12-20 eggs every 7 to 10 days for a period of several months. If a pair is pulsing its reproductive output in this way, the breeding tank will soon contain young of several different sizes, since older fry are tolerated in the presence of younger siblings. Feed the fry on microworms for their first few days of mobile life and then on brineshrimp nauplii. Fry survival is enhanced in an established aquarium containing a mature sponge filter. The young are sexually mature 14 months after spawning, at 7.5cm(3in) for males, slightly less for females.

The second large representative of the genus, *J. regani*, has the lateral stripes of *J. marlieri* but lacks the vertical bars. The maintenance requirements and reproductive patterns of the two species are identical. The closely related *Chalinochromis* species respond to the same care.

Pairs of the three dwarves of the genus, *J. dickfeldi*, *J. ornatus* and *J. transcriptus*, as well as dwarf species such as *Telmatochromis bifrenatus* and *T. vittatus*, will thrive in tanks 60cm(24in) long. All these species are about 7.5cm(3in) long and require the same care as *J. marlieri* in captivity.

Lamprologus brichardi

● **Habitat:** Large schools of this midwater-swimming species live in close association with rock faces that drop off steeply into the lake's depths. Though individuals can be found as deep as 30m(98ft), most reproductive activity takes place much nearer to the surface.

● **Length:** Males up to 9cm(3.5in); females slightly smaller.

● **Sexing:** Females are smaller than males and have a less rounded cranial profile. Length of fin filaments is not a reliable sexual distinction.

● **Feeding type:** Micropredator.

● **Compatibility:** Aggressive towards own species other than members of its extended 'family' in captivity, as well as towards other *Lamprologus* of a similar appearance. Otherwise, sexually inactive individuals make good community residents with companions too large to be easily swallowed. Parental fish defend a territory about 45cm(18in) square and are quite capable of killing tankmates unwilling, or unable, to respect its limits.

The easiest way to secure a pair is to raise a group of young to maturity together. It is not unusual for this normally monogamous species to shift to harem polygyny if a surplus of females is present. Communal spawning – which occurs frequently in nature – is also likely if siblings are reared to maturity in a large, well-stocked community setting, where a single pair might encounter difficulties establishing a breeding territory. Sexual bonds tend to be stable even in the absence of target fish. Spawns can number 200 ovoid, olive-green eggs, but clutches of 50-75 are more normal. Fry are permitted to remain within their parents' territory, where they assist in the defence of their younger siblings, until they are sexually mature, eight months after spawning, at about 4cm(1.6in).

An albino form of *L. brichardi* is commercially available and several distinctive *L. brichardi*-like fishes have been exported from different regions of the lake. Whether the 'black-faced', 'daffodil' and 'Kasagere' *Lamprologus* represent geographic races or subspecies of *L. brichardi*, or distinct species, remains to be determined. Take care to prevent hybridization between these attractive cichlids. *L. pulcher* is a superficially similar species with more pronounced rusty orange lateral spots and fin markings. More aggressive than *L. brichardi*, but shares the features of its reproductive behaviour.

Below: **Lamprologus brichardi** *Elegant flowing finnage.*

Lamprologus calvus

Pearly compressiceps

● **Habitat:** Found in close association with rocky bottoms to a depth of 12m(40ft). All known collecting sites are along the lake's southern and eastern shores.

● **Length:** Males up to 13cm(5in); females up to 9cm(3.5in).

● **Sexing:** Males grow larger and have a more massive head, as well as deeper dorsal and anal fins.

● **Feeding type:** Micropredator.

● **Compatibility:** Will eat fish as large as a female guppy, but poses no threat to tankmates too large to swallow. However, it is intolerant of others of its own species and of the superficially similar *L. compressiceps,* but usually gets along well with other retiring Tanganyikan cichlids. A slow-moving fish, easily intimidated by more assertive tankmates in a community tank, *Lamprologus calvus* is often the loser at feeding time. Digs less than other representatives of this group.

Like *L. leleupi* (page 85), males of this species will practise harem polygyny in captivity when the opportunity arises. Females prefer to deposit their eggs on the vertical wall of a cleft in the rockwork, but will accept a flowerpot or similar enclosed space as a spawning site in captivity. Spawns can number up to 100 whitish eggs, but half that number is closer to the norm. A significant percentage of the eggs often fails to develop. Whether this is the result of male failure to fertilize the entire clutch, or the female's rather casual hygienic behaviour towards the zygotes is unclear. Growth is slow, but *L. calvus* begins breeding well before it attains its maximum size. The young are sexually mature at about 12 months after spawning, at 5.7cm(2.25in) for males, 3.75cm(1.5in) for females.

The closely related *L. compressiceps* is somewhat deeper bodied than *L. calvus* and lacks its profusion of white dots. However, the maintenance requirements and reproductive pattern of these two species do not differ significantly.

Below: **Lamprologus calvus** *An inept predator on smaller fishes.*

Above: **Lamprologus leleupi** *Brilliantly coloured, medium-sized cichlid.*

Lamprologus leleupi

● **Habitat:** Found in close association with the bottom over rocky shores to a depth of 70m(230ft). Relatively uncommon in water less than 40m(130ft) deep.

● **Length:** Males up to 10cm(4in); females up to 8cm(3.2in).

● **Sexing:** Females are smaller than males, with slightly shorter vertical fins and ventrals.

● **Feeding type:** Micropredator. Offer colour food regularly.

● **Compatibility:** Extremely aggressive towards its own species. Do not keep more than a single pair in tanks less than 150cm(60in) long. Will prey on tankmates the size of a male guppy. Sexually inactive individuals are otherwise good neighbours in a Tanganyikan community tank. Parental fish behave in the same manner as *L. brichardi* (see page 83).

Pairs breed readily if handled in the manner recommended for monogamous cichlids. In a large tank, a single male will usually spawn with all the available females, but generally involves himself in caring only for the last brood of fry he has sired. Spawns can number up to 200 greenish white, ovoid eggs. Unlike *L. brichardi*, adult fish do not tolerate the presence of older fry in their territory with the onset of another bout of reproductive activity, so be sure to separate parents and offspring six to eight weeks after spawning. The young grow more slowly and less evenly than do *L. brichardi* fry, so sort them by size to prevent sibling cannibalism. Sexual maturity is attained 18 months after spawning, at a length of 6.5cm(2.5in) for males, slightly less for females.

A subspecies, *Lamprologus l. longior*, is often commercially available. It is characterized by its intense, constant yellow-orange coloration and somewhat slenderer build. *L. l. leleupi* is a somewhat stockier, golden ochre fish that turns intense lemon yellow when sexually active. It has been largely displaced by *L.l. longior*.

Above: **Lamprologus sexfasciatus** *A strikingly marked species.*

Lamprologus sexfasciatus

● **Habitat:** Found in close association with rocky bottoms to depths of 5m(16ft).
● **Length:** Males up to 14cm(5.5in); females slightly smaller.
● **Sexing:** Females are somewhat smaller and fuller-bodied than males, but the only reliable means of sexing this species is by direct examination of the genital papillae.
● **Feeding type:** Micropredator.
● **Compatibility:** As described for *L. leleupi*, but given its larger adult size, *L. sexfasciatus* requires roomier quarters. A single pair will live happily in an aquarium 90-120cm(36-48in) long.

Handle this species in the manner recommended for *L. leleupi* (see page 85). It is not unusual for a pair of this slow-maturing species to go through several false spawns before actually producing a clutch of several hundred ovoid, off-white eggs. Although they appear infertile, the eggs are perfectly viable. Offer the very small fry microworms for the first few days,

then *Artemia* nauplii. Fry survival seems greater if a mature sponge filter is present in the breeding tank for the fry to browse over. Remove the young from the breeding tank by the sixth week after spawning. They reach sexual maturity after about 18 months, at 9cm(3.5in) for males, 7.5cm(3in) for females.

Lamprologus signatus

● **Habitat:** Lives in close association with empty shells of the snail *Neothauma*, which occur in extensive drifts on open, sandy bottoms at depths of 10-100m(33-330ft).
● **Length:** Males up to 5cm(2in); females up to 3cm(1.2in).
● **Sexing:** Males grow much larger than females and have deeper and more strikingly marked vertical fins.
● **Feeding type:** Micropredator.
● **Compatibility:** Males defend a territory about 30cm(12in) square against others of their sex. Several males can coexist as long as the tank is large enough to afford each a territory. As with most harem polygynists, females defend a much smaller area, restricted to

the vicinity of their shell. This species tolerates other Tanganyikan cichlids of the same size, or slightly larger, outside periods of sexual activity. However, parental fish become extremely intolerant of other bottom-dwelling fish. Easily bullied by larger lamprologines. Does best when kept exclusively with midwater-swimming tankmates too large to be swallowed.

This species will thrive equally well whether maintained in pairs or harems. Sexual bonding occurs freely, any mature individual spawning with another of the opposite sex but, unlike most lamprologines, female shell-dwellers display actively to the male before spawning. *L. signatus* insists on an empty snail shell as a spawning site. Eggs are placed deep within the shell, well out of sight, so female reluctance to leave its interior following a bout of courtship is the only indication that spawning has occurred. Up to 100 tiny fry emerge from the shell a week later and are allowed to forage near the opening under their mother's supervision. Offer them infusorians and microworms during the first week. Fry survival is always greater in a well-established aquarium containing a mature sponge filter. The female herds the fry back into the shell at night and the male blocks its aperture with sand and mounts guard over his immured family until morning, when he removes the plug and moves off to provide perimeter defence. The female continues to protect the fry for up to three weeks. She then chases them from her territory and eats any that remain behind. This can provoke an attack by the male, whose protective tendencies persist longer than the female's. To avoid intersexual conflict, separate the parents from their offspring at this point. Even at this early age, males are 50 percent larger than females. The young reach sexual maturity 10 months after spawning, at 4cm(1.6in) for males, 2cm(0.8in) for females.

Below: **Lamprologus signatus** ♂ *An appealing little shell-dweller.*

Tanganyikan mouthbrooding cichlids

Although the species in this category stem from several different evolutionary lineages, they are for the most part maternal mouthbrooders (i.e. they incubate their eggs in the throat sac) with an openly polygamous mating system and should be treated accordingly (see page 57 of the section dealing with *Breeding and rearing*). The goby cichlids (*Eretmodus*, *Spathodus* and *Tanganicodus*) are monogamous mouthbrooders and both sexes share the task of incubation. They require the same care as any other monogamous cichlid (see page 54 for advice on breeding these cichlids).

Cyathopharynx furcifer

● **Habitat:** Schools of this social species are found at depths of 9-12m(30-40ft) just off rocky slopes.
● **Length:** Males up to 20cm(8in); females up to 16.5cm(6.5in).
● **Sexing:** Territorial males are instantly recognizable by their scintillating blue coloration. Subordinate males differ from females in their duskier overall coloration and longer ventral fins.
● **Feeding type:** Micropredator.
● **Compatibility:** Despite its rather large size, this species is quite unaggressive towards both its own species and other fishes. It is possible to keep several males in a tank at least 180cm(72in) long. However, even in an aquarium of this size only one male will ever display its magnificent courting dress. Easily bullied by smaller, more assertive cichlids, this species is best housed alone or in the company of midwater-swimming, non-cichlid companions. Extensive pit digging by the dominant male makes it difficult to keep any but potted plants in its tank.

In nature, males build an elaborate crater nest by carrying sand to the top of rocky outcrops. In captivity, they simply excavate the largest pit possible at one end of the tank.

Below: **Ophthalmochromis nasutus** *One of the delicate featherfins.*

Courtship is prolonged; the male darts out from his nest, displays to the female and darts back again. Eventually she follows him to the pit and deposits a few eggs, which she immediately picks up. She then mouths the light tips of the male's ventral fins, which are folded back to the genital papilla during spawning. Clutches of up to 40 eggs have been reported, but two dozen is more usual. At 25°C(77°F), the incubation period is 21 days. Females often prove unreliable mothers and it may be helpful to offer them more privacy; if this does not have the desired effect, the only option may be to incubate the zygotes artificially. The newly released fry can take brineshrimp nauplii for their initial meal. They share the sensitivity of *Xenotilapia ochrogenys* to nitrogen cycle mismanagement and need similar care. The young begin breeding 10-12 months after spawning, at a length of 10cm(4in) for males, 9cm(3.5in) for females. However, up to 95 percent of the eggs fertilized by such young males prove infertile. Males do not attain full reproductive competence until they are at least 18 months old.

Four different geographic colour forms of *C. furcifer* are known, and fishkeepers should prevent hybridization between them. All the Tanganyikan featherfins, which consist of the closely related genera *Aulonacranus*, *Cardiopharynx*, *Cyathopharynx* and *Ophthalmotilapia*, require the same care in captivity. These delicate species are best attempted by more experienced cichlid keepers.

Cyphotilapia frontosa

● **Habitat:** This species is restricted to rocky habitats. Juveniles are found at depths between 18 and 50m(60-164ft), but adults are rarely encountered in water less than 27m(90ft) deep.
● **Length:** Males up to 35cm(14in); females up to 25cm(10in).
● **Sexing:** Males are larger, with long, flowing fins and a very well developed nuchal hump.
● **Feeding type:** Piscivore.
● **Compatibility:** Males are extremely intolerant of one another. Do not house more than a

Below: **Cyphotilapia frontosa** *Keep in a spacious aquarium.*

single male per tank, even in aquariums more than 180cm(72in) long. A male is also apt to injure females if the fish are maintained in crowded quarters. A 150cm(60in) – long tank is the absolute minimum for a trio of *C. frontosa*. Tankmates that are too large to be easily swallowed are ignored. Despite its large adult size, this slow-moving species is easily intimidated by smaller, more aggressive cichlids, such as mid-sized *Lamprologus*. It does best when kept with other Tanganyikan mouthbrooders of a similar temperament. It neither digs nor molests rooted plants.

Well-conditioned fish spawn readily if managed like any other openly polygamous cichlid. After a rather desultory courtship, the female, followed closely by the male, deposits a row of eggs on a solid surface. Only after the male has fertilized them does the female pick up the eggs. Well-fed females can produce as many as 60 large, off-white eggs in a spawning, but 20-25 is closer to the norm. The incubation period is 28 days at 27°C(80°F). This species does not care for its fry once they are released. Young females have a reputation for maternal unreliability and even some experienced females consistently devour their clutches. If isolating the egglaying female fails to correct this problem, the only alternative is to use an 'artificial mouth' to bring the zygotes to term. The robust young can manage brineshrimp nauplii, sifted *Daphnia* and powdered prepared foods immediately after emerging from the female's mouth. With good feeding and frequent small water changes, they grow rapidly and reach sexual maturity 10 or 12 months after release, at 15cm(6in) for males, 10cm(4in) for females.

Cyprichromis nigripinnis

● **Habitat:** A pelagic species found offshore on rocky slopes over a depth range of 10-50m(33-164ft).
● **Length:** Males up to 10cm(4in); females up to 9cm(3.5in).
● **Sexing:** Females are smaller than males and lack their dusky, iridescent blue-edged vertical fins.
● **Feeding type:** Micropredator.
● **Compatibility:** Always keep this extremely gregarious species in groups of six or more in captivity. Male aggression is highly ritualized, so it is usually possible to maintain several males together if they are not crowded. The fishes require plenty of swimming space; a 150cm(60in) – long tank is the minimum recommended for any *Cyprichromis* species. Easily bullied by more assertive companions, but the smaller *Julidochromis* species and the several genera of goby cichlids make good tankmates, as do *Callochromis* and *Xenotilapia*. Neither digs nor molests rooted plants. Keep this accomplished jumper in a tightly covered tank.

Below: **Cyprichromis nigripinnis** ♂ *A pelagic 'sardine cichlid'.*

Above: **Tanganicodus irsacae** *Smallest Tanganyikan goby cichlid.*

Males actively court ripe females and contend vigorously for their attention. This is the only time that their aggressive behaviour may lead to injuries. After reciprocal circling, the female expels an egg and immediately dives down to seize it. The male then tilts slightly to one side and allows the female to mouth his vent. Spawns rarely exceed 20 eggs and are usually smaller. Egg-carrying females do not always display a bulging throat, but can be identified from above by their slightly outspread gill covers. In a single-species breeding tank, it is not necessary to isolate the female, since adults ignore newly released fry. The incubation period lasts 21 days at 25°C(77°F). The fry are large enough to take brineshrimp nauplii immediately on release. Because of their short intestinal tract and extremely rapid metabolism, they require numerous small daily feeds to prosper. They are also extremely sensitive to dissolved metabolites. The young attain sexual maturity between 10 and 12 months after spawning, at 7cm(2.75in) for males, 5.7cm(2.25in) for females. *C. microlepidotus* and *C. leptasoma* require the same care in captivity.

Tanganicodus irsacae

● **Habitat:** This bottom-living species is found in the surge zone of cobble beaches. Like other Tanganyikan surf-dwellers, it lacks a functional swimbladder and moves over the bottom with droll hopping motions.

● **Length:** Males up to 9cm(3.5in); females slightly less.

● **Sexing:** Apart from the marginal difference in size, there is no way of distinguishing between the sexes.

● **Feeding type:** Micropredator. Swims with difficulty; provide foods that sink to bottom.

● **Compatibility:** Pairs are very aggressive towards companions of their own species and have little use for the close company of other goby cichlids. Despite its small adult size, it is not possible to keep more than a single pair in a tank 120cm(48in) long. This species is easily bullied by larger *Lamprologus* species, but otherwise gets along well with a wide range of Tanganyikan cichlid companions. Like other goby cichlids, it poses no threat even to much smaller midwater-swimming tankmates. Not a digger, but a skilful jumper; provide a tightly covered tank.

The best way to secure a pair is to rear a group of young to maturity together and, as soon as two fish begin to act like a pair, remove the remaining specimens. Courtship is desultory and followed by reciprocal circling, during which the eggs are deposited and fertilized. One of the spawning fish picks up the flattened, discoid eggs and carries them for half of the incubation period. Then its partner takes over the task of brooding the zygotes. Spawns range from 6 to 20 eggs. Because so few eggs do not produce a conspicuous distension of the throat and the egglaying parent continues to feed during the incubation period, it is easy to miss a spawning. Often the first indication that breeding has occurred is the appearance of fry among the tank's rockwork. At 27°C(80°F), the young are released 21 days after spawning. There is no parental care once the fry have been released; in fact their parents ignore them. The young can take brineshrimp nauplii and powdered prepared foods for their initial meal. They are extremely sensitive to any build-up of metabolic waste. However, if kept under a regime of frequent small partial water changes, they are not difficult to rear. It takes the young a year to attain full adult size, and an additional 12 months to attain sexual maturity.

Tropheus moorii
● **Habitat:** Found over rocks supporting dense algal growth. Adults are seldom encountered at depths greater than 10m(33ft).
● **Length:** Males up to 14cm(5.5in); females up to 10cm(4in).
● **Sexing:** Males are larger than females and sport a distinctly 'Roman' nose, but direct examination of the genital papillae is the only reliable means of sexing this species.
● **Feeding type:** Herbivore.

Below: **Tropheus moorii** *Available in many colour forms.*

● **Compatibility:** Males are extremely intolerant of one another. Avoid introducing mbuna tankmates, but smaller lamprologines and midwater-swimming companions are tolerated. To thrive in captivity, a group of *T. moorii* require a tank at least 120cm(48in) long. This species will eat soft-leaved aquatic plants, but does not dig.

Like other openly polygamous cichlids, this species spawns readily when maintained in single male/multiple female groups. The male does not display to a ripe female from a fixed territory, but follows her about the tank, actively soliciting her attentions. The pair drop to the bottom and start to circle one another. The female expels the large eggs one at a time and immediately picks them up. They are fertilized in her mouth as she nips directly at the male's vent. Broods of up to 20 fry have been reported in captivity, but a dozen eggs per spawning is closer to the norm. The incubation period is 28 days at 27°C(80°F). Well-

developed brood care is characteristic of all *Tropheus* species and females exercise the greatest caution before releasing their progeny for the first time. This may account for observations that the incubation period lasts 30 days in *Tropheus* species. If the breeding group is accommodated in its own tank, there is no need to separate the parental female from her companions, since adequately fed adults ignore newly released fry. The robust fry can immediately take brineshrimp nauplii and powdered foods and quickly emulate the grazing behaviour of their elders. With due attention to good aquarium hygiene, they are easy to rear and grow rapidly. The young are sexually mature between 10 and 12 months after spawning, at 7.5cm(3in) for males, slightly less for females.

There are over two dozen recognized geographical colour forms of this species. Some of these refuse to interbreed in captivity and may actually constitute distinct biological species. As a general rule, you should make every effort to avoid crossing these colour forms.

Xenotilapia ochrogenys
● **Habitat:** Lives in large schools over open, sandy substrates at depths not exceeding 20m(66ft). Breeding groups usually inhabit much shallower water, no more than 4m(13ft) deep.
● **Length:** Males up to 15cm(6in); females 10cm(4in).
● **Sexing:** Males are easy to distinguish by their dazzling colours and breeding dress.
● **Feeding type:** Micropredator.
● **Compatibility:** Males defend a territory about 60cm(24in) square against other males, but several males will coexist if the tank is large enough to afford each a territory. Although easily bullied by more aggressive cichlids, *X. ochrogenys* does get along well with *Cyprichromis* species or any other midwater-swimming fishes. This species does not eat aquatic plants, but its habit of moving

Above: **Xenotilapia ochrogenys** ♂ ♀ *A courting pair (male below).*

massive amounts of gravel when digging spawning pits places all save potted specimens in continual jeopardy.

Xenotilapia ochrogenys spawns freely if kept in multiple female groups in a tank of its own. Spawning is preceded by vigorous courtship of the female and intense pit digging by the male. The female deposits and picks up a few eggs after a period of reciprocal circling and vent nudging. Clutches range from 5 to 30 eggs. It is easy to overlook a spawning; such small clutches do not produce a marked throat bulge and females will continue to feed, albeit at a reduced rate, while carrying them. Young females have a reputation for maternal unreliability, and even older individuals may abort the incubation sequence if the breeding tank is too small for them to avoid the male's attentions. At 27°C(80°F) the zygotes are carried for 21 days. The newly released fry can take brineshrimp nauplii and fine powdered food for their initial meal. They are extremely sensitive to accidental pollution of their tank and punctilious attention to proper nitrogen cycle management is

vital. In order to avoid damaging fights between maturing males, it is a good idea to segregate the sexes as the young begin to manifest sexual colour differences. This is possible at six to eight months after release, but the young fish do not attain reproductive competence until they are a year old, at 7.5cm(3in) for males, and slightly smaller for females. It takes them an additional year to grow to their full adult size.

This dazzling species is representative of the many substrate-sifting cichlids of the genera *Callochromis* and *Xenotilapia*. Their scintillating colours and reasonably mellow dispositions make them an obvious choice for the fishkeeper, but until recently they have enjoyed limited availability. Adults are very difficult to ship successfully because of their extreme sensitivity to dissolved metabolites and their high dissolved oxygen requirements. Juveniles are much better travellers, so with the increasing availability of tank-reared fry, these species can be expected to gain the popularity they deserve.

LAKE MALAWI CICHLIDS

Like their Tanganyikan counterparts, these fishes are all acutely intolerant of dissolved metabolites. Efficient biological filtration and frequent partial water changes are absolutely essential to their well-being in captivity. To minimize aggression under aquarium conditions, maintain these cichlids at relatively high population densities, but high stocking rates are practical only when scrupulous attention is paid to nitrogen cycle management. Malawian cichlids require hard, alkaline water to prosper. A temperature range of 22-25°C(72-77°F) will suffice for day-to-day maintenance, with an increase to 27-30°C(80-86°F) for breeding. These cichlids are capable jumpers, so keep them in a tightly covered tank. Most mbuna eat soft-leaved plants, while the males of the majority of Malawian cichlids move enormous volumes of gravel when constructing a nest pit.

These fish are all openly polygamous maternal mouthbrooders. Managed correctly, they breed readily in captivity. Spawning follows a prolonged and often vigorous courtship. After a period of reciprocal circling, the fish spawn in the typical haplochromine manner. Although the eggs are fertilized within the female's mouth, she does not elicit ejaculation by nipping at the pseudo-ocelli often present on the male's anal fin. Her target is the male's vent, where his genital papilla often contrasts strongly with the background colour of the abdominal region. At 30°C(86°F), the incubation period lasts 21 days. The robust fry can take brineshrimp nauplii and powdered food for their initial meal and are easy to rear. When selecting future breeding stock remember that all male haplochromines grow more quickly than females.

Sexually active male Malawian cichlids are strongly territorial. In captivity, favourable living conditions promote year-round male sexual activity, while space limitations preclude the natural expression of typical territorial behaviour. This is why, with rare exceptions, it is not possible to house several males of the same species together in the presence of females.

The unavoidable crowding of these fish under aquarium conditions also appears to explain the phenomenon of hyperdominance, in which one male will harass males of other species so severely that he prevents them from expressing any sort of territorial or courtship behaviour. In the absence of appropriate sexual cues from their own males, females of other species eventually respond positively to the hyperdominant male's courtship, producing unwanted but often viable hybrids.

Ideally, Malawian cichlids are best kept in 'harems' of a single male and several females, one species to a tank. In practice, however, most aquarists prefer to maintain a mixed-species community of cichlids. To reduce the risk of unwanted hybridization, observe the following guidelines:

● Do not house together species with markedly different aggressive tendencies. The male of the most aggressive species will become hyperdominant.

● Do not house together very closely related species, or those with very similar male courting dress. Females are less likely to succumb to the overtures of a very differently coloured male of another species.

● Do not house together species whose females are very similar in appearance. Fry initially resemble their mother, not their father. The results of hybridization between two species with essentially identical female colours will not be recognized for what they are until males begin to colour up. However, if the females of the hybridizing species have very different colour patterns, hybrid fry are easy to recognize.

Hyperdominance is chiefly a consequence of limited living space. The larger the tank available for a community of Malawian cichlids, the less the likelihood of unwanted hybridization. If these guidelines are followed, most of the popular Malawian species in the 7.5-15cm(3-6in) range can be housed in tanks 150-180cm(60-72in) long.

Mbuna compatibility

Group 1
Species that will behave aggressively towards other fish in tanks more than 180cm (72in) long and are likely to kill or injure tankmates in smaller aquariums.

Genyochromis mento
Melanochromis chipoke
M. loriae
M. melanopterus
M. simulans
M. 'lepidophage'
All Petrotilapia spp.

Pseudotropheus brevis
Ps. crabro complex
Ps. elongatus complex
Ps. hajomaylandi
Ps. lucerna
Ps. tursiops
Ps. williamsi complex

Group 2
Species that usually forgo serious aggression in tanks more than 180cm(72in) long; large specimens are likely to injure or kill other fish in tanks less than 150cm(60in) long.

Labeotropheus fuelleborni
Melanochromis auratus
M. parallelus
M. vermivorous
Pseudotropheus aurora

Ps. greshakei
Ps. heteropictus
Ps. cf. livingstonii
Ps. lombardoi
Ps. zebra complex

Group 3
Species that usually forgo serious aggression in tanks more than 150cm(60in) long; large specimens can pose some risk to their companions in tanks less than 120cm(48in) long.

Cyathochromis obliquidens
All Gephyrochromis spp.
Labeotropheus trewavasae
Melanochromis interruptus
M. johanni complex
Pseudotropheus barlowi
Ps. elegans

Ps. gracilior
Ps. lanisticola
Ps. macrophthalmus complex
Ps. cf. microstoma
Ps. minutus
Ps. socolofi complex
Ps. tropheops complex

Group 4
Species that usually forgo serious aggression in tanks more than 120cm(48in) long.

All Cynotilapia spp.
Iodotropheus sprengerae

All Labidochromis spp.

Mbuna

The Tonga people, who live on the western coast of Lake Malawi, apply the term 'mbuna' (rockfish) to representatives of a distinctive group of haplochromines defined by a number of morphological and behavioural characteristics. It is easy to see why the mbuna were the first cichlids exported as aquarium fish. Most species range from 6.5-13cm(2.5-5in), have dazzling male and, in some instances, female coloration, are vivacious in manner and easy to maintain. They have been compared to marine fishes, such as the damselfishes and wrasses. The comparison is apt, for mbuna are both brightly coloured and aggressive aquarium residents. To keep them successfully in captivity it is important to understand the basis of their aggressive behaviour.

As with most other cichlids, mbuna aggression is related to the defence of a territory and this territoriality is usually a function of reproduction. Mbuna differ in that males of many species defend a multipurpose territory that provides food, shelter from predators and a spawning site. In a few instances, even females defend a feeding territory against potential competitors of the same and other species. Obviously, the value of such territories persists well beyond the duration of a single

spawning episode, hence the phenomenon of persistent territorial defence in nature.

The size of the area defended by a resident fish in nature is so large that in captivity it is seldom practical to think of housing more than a single male of a given species per tank, even in well-furnished aquariums as large as 180cm(72in) long. Matters are further complicated by the fact that a territory holder will attack a similarly coloured species, or one with a very similar feeding pattern, just as vigorously as an intruder of the same species. However, it is possible to avoid serious aggression in a mbuna tank by carefully selecting tankmates that do not perceive one another as threatening.

The table divides the most widely available mbuna species into four groups in decreasing order of their aggressiveness under aquarium conditions. While making allowances for the individuality of cichlids in general, it should still be possible to assemble a reasonably harmonious community of these cichlids, using these guidelines.

● Representatives of adjacent groups, e.g., groups 2 and 3, are more likely to coexist separated than those in disconnected groups, e.g., 1 and 3.

● Among cichlids in the same group, fishes of the same genus are more likely to prove incompatible than representatives of different genera.

● Among representatives of the same genus, species with similar body shapes are more likely to prove incompatible than those that differ in this regard.

● Regardless of the group to which they belong, species with similar male courting coloration are apt to prove incompatible when housed in the same aquarium.

Below: **Pseudotropheus zebra**
The striking coloration of the male has assured its popularity.

Mixing mbuna with other Malawian haplochromines is not desirable, nor should you house mbuna with their Tanganyikan cousins – *Petrochromis*, *Simochromis* and *Tropheus* species – since the latter will not prosper in their company. Well-fed mbuna typically ignore midwater-living companions too large to make a convenient mouthful, but their own vivacity usually renders dither fish superfluous in their tank.

Mbuna breed freely in a community setting. With the exception of the long-snouted *Melanochromis* species, which prey efficiently on newly released cichlid fry, adults are disinclined to molest either their own young or those of other species. Assuming their aquarium is well furnished with rockwork and free of potential predators, mbuna fry have a good chance of growing to maturity in a community of their elders, as long as they are fed. All mbuna thrive on the diet recommended for herbivorous fishes. Do not offer them *Tubifex* worms in any form or frozen brineshrimp. The coloration of many species responds to colour foods.

Mbuna species were originally defined on the basis of morphological characteristics. However, each isolated stretch of shoreline or offshore island group supports its own distinctive colour phenotypes and ichthyologists disagree over whether these should be treated as regional colour variants of a single species, such as *Pseudotropheus tropheops*, or as valid, albeit closely related, species. Until more is known about the genetics and behaviour of these often very distinctive phenotypes, aquarists are well advised to treat them as distinct species and to avoid crossing them.

Iodotropheus sprengerae

Rusty cichlid

● **Habitat:** Found over a wide range of rocky habitats and sometimes even in the intermediate zone at Boadzulu, Chinyamwezi and Chinyamkwazi Islands. Although individuals have been recorded in water 40m(131ft) deep, this species is most abundant at depths of 3-15m(10-50ft).

● **Length:** Males up to 9cm(3.5in); females up to 7cm(2.75in).

● **Sexing:** Both sexes are the same colour and can be distinguished only by the presence of large, clearly defined pseudo-ocelli on the male's somewhat longer anal fin.

● **Feeding type:** Micropredator.

● **Compatibility:** Males are territorial for a brief period before spawning, and then only towards one another. It is thus possible to keep several males in a tank more than 120cm(48in) long. Do not house *I. sprengerae* with more belligerent mbuna.

Below: **Iodotropheus sprengerae** ♂ *Hardy and unaggressive.*

Above: **Labeotropheus fuelleborni** *One of many colour forms.*

Spawns range from 5 to 60 eggs. This species does not undergo a colour metamorphosis and displays remarkable reproductive precocity. The young can begin breeding as early as the 14th week after spawning, at 3.2cm(1.25in) for males, 2.5cm(1in) for females. They do not attain full adult size until they are at least 8-10 months.

The only species with which the rusty cichlid is likely to be confused is *Labidochromis vellicans*. Both species share a rusty base colour, but *L. vellicans* has a more pointed snout. By virtue of its relatively mellow temperament, attractive coloration and willingness to breed in captivity, *Iodotropheus sprengerae* is the ideal mbuna for a beginner to keep.

Labeotropheus fuelleborni

● **Habitat:** Abundant over areas of broken rock and in the intermediate zone throughout the lake. Most abundant in water less than 5m(16.5ft) deep, although individuals can penetrate as deep as 18m(60ft).
● **Length:** Males up to 12cm(4.7in); females slightly smaller. Captive specimens often reach 15cm(6in).
● **Sexing:** Males grow larger, and their colour metamorphosis begins at 6.5cm(2.5in).

● **Feeding type:** Herbivore. Appreciates fresh vegetables.
● **Compatibility:** Males are extremely aggressive towards males of their own species but tend to disregard males of other species. A significant exception to this rule are males of *L. trewavasae*. Indeed, efforts to house these very similar species together in captivity usually end badly for one or the other. Very large males are apt to become hyperdominant in tanks 180cm(72in) long.

Males are likely to harass egg-carrying females, whose spawns can number 75 eggs. The onset of sexual activity occurs between 9 and 12 months after release, at 7.5cm(3in) for males, slightly less for females.

Like *Ps. zebra*, this species is characterized by both colour polymorphism and extensive geographic colour variation. Over 20 distinct races of *L. fuelleborni* have been reported in nature and several of these have been exported. Fortunately for aquarists, the available evidence argues against the existence of cryptic biospecies (i.e. biologically distinct fishes that do not differ in shape, but often differ in details of coloration) of the sort that make up the *Ps. zebra* complex. True-breeding strains of the orange/black morph are also available.

Above: **Labeotropheus trewavasae** *Offer fresh vegetable foods.*

Labeotropheus trewavasae

● **Habitat:** Most common over large rocks, but has been recorded from other habitats throughout the lake. Individuals are fairly evenly distributed to depths of 20m(66ft).
● **Length:** Males up to 12cm(4.7in); females slightly smaller. Captive specimens often reach 15cm(6in).
● **Sexing:** Males begin to change colour at 6.5cm(2.5in).
● **Feeding type:** Herbivore.
● **Compatibility:** Males are less markedly territorial than *L. fuelleborni*, but do not house the two species together (see *L. fuelleborni* page 99). *L. trewavasae* seldom displays hyperdominance in captivity.

Sexual maturity is reached between 9 and 12 months after release, at 7.5cm(3in) for males, slightly less for females. Spawns range from 10 to 60 eggs.

L. trewavasae is characterized by both colour polymorphism and extensive geographic colour variation. Several of the 13 recognized races have been exported. Of these, the most striking is the Chitende population, whose males are rusty ochre with icy mauve vertical fins. The Tumbi Island race is characterized by both orange/black and solid orange female morphs. A strain in which both sexes are orange and true-breeding orange/black strains are commercially available.

Labidochromis zebroides

Likoma Island clown
● **Habitat:** The interstices of large and medium-sized rocks at Likoma Island. This shallow-water species is always found at depths of less than 6m(20ft).
● **Length:** Males up to 6.5cm(2.5in); females up to 5cm(2in).
● **Sexing:** Females are a uniform silvery grey. Males begin to change colour at 2.5cm(1in).
● **Feeding type:** Micropredator.
● **Compatibility:** The behaviour of this species is typical of the genus as a whole. Males defend small territories, about 30cm(12in) square, based upon a cave or similar shelter. Given their small adult size and modest space requirements, it is possible to house several males in a tank more than 120cm(48in) long. Despite its small adult size, this species can be safely housed with larger mbuna, as long as the tank is well furnished and contains plenty of cover. This species does not manifest hyperdominant behaviour in captivity.

Sexual maturity coincides with the complete expression of male coloration, at about eight months after release, at 3.8cm(1.5in) for males, slightly less for females. Spawns range from 5 to 30 eggs. Unlike other mbuna species, female *Labidochromis* do not practise parental care of their fry.

Above: **Labidochromis zebroides** ♂ *A most beautiful mbuna.*

Labidochromis freibergi is a similarly coloured species from the Likoma Island group and is often available commercially. It has a shorter snout, more rounded cranial profile and deeper body. For no immediately obvious reason, *L. freibergi* is frequently sold under the name *Pseudotropheus minutus.*

Melanochromis johanni

● **Habitat:** Occurs over rocky habitats and in the intermediate zone along the eastern shoreline of the lake. This is a shallow water species; its distribution is restricted to water 6m(20ft) deep or less.

● **Length:** Males up to 9cm(3.5in); females slightly smaller. In captivity, males can grow up to 12.7cm(5in); females up to 10cm(4in).

● **Sexing:** In some strains, males begin to change colour at 2.5cm(1in).

● **Feeding type:** Omnivore.

● **Compatibility:** Young males tolerate each other well enough, but as they grow larger, it is virtually impossible to keep more than one male per tank. Best kept with other *Melanochromis* species, for its small adult size puts it at a severe disadvantage. This species seldom displays hyperdominance in captivity.

Below: **Melanochromis johanni** ♂ *A territorial male.*

Sexual maturity is attained at about eight months after release, at 6.5cm(2.5in) for males, 5cm(2in) for females. Spawns range from 7 to 60 eggs.

Two populations with identical juvenile coloration but different male and, in one instance, female colour patterns have been exported. Instead of the parallel stripes of *M. johanni*, the males of the Chisumulu Islands, are sooty black with two parallel rows of bright blue blotches on the flanks. The second population, restricted to Likoma Islands, is characterized by females marked with black and blue-white stripes. Until their status is resolved, it is best to treat both as distinct species.

Pseudotropheus lombardoi

● **Habitat:** Only found at Mbenji Island and most abundant at the rock/sand interface. Also occurs over large, flat rocks. Lives in water 2-25m(6.6-82ft) deep, but most abundant at depths greater than 10m(33ft).
● **Length:** Males up to 10cm(4in); females slightly smaller. Like most mbuna, this species grows substantially larger in captivity.
● **Sexing:** Males begin to change their bright blue juvenile coloration to the yellow adult coloration at 4cm(1.6in).

● **Feeding type:** Herbivore.
● **Compatibility:** Males are less belligerent towards other mbuna than *Ps. zebra* (see page 105).

Male sexual maturity coincides with the completion of the colour metamorphosis, eight months after release, at 6cm(2.25cm). Females mature concurrently, at 5cm(2in).

Spawns range from 6 to 75 eggs, depending upon the size and condition of the female. These often lose their bright blue juvenile coloration after carrying several broods and become a pale beige with a blue wash on the flanks.

Pseudotropheus macrophthalmus 'yellow head'

● **Habitat:** Particularly common along rock/sand interfaces at Likoma and Chisumulu Islands. Specimens have been recorded as deep as 23m(75ft), but most abundant in the shallows, between 1-5m(3.2-16.5ft).
● **Length:** Males up to 11.5cm(4.5in); females up to 9cm(3.5in).
● **Sexing:** Females are silvery grey with a lilac wash and a row of small dark blotches along the midlateral line and the back. Distinctive male coloration begins to appear at 5cm(2in).

Below: **Pseudotropheus lombardoi** ♂ *Striking male coloration.*

Above: **Pseudotropheus macrophthalmus 'yellow head'** *Male.*

● **Feeding type:** Herbivore.
● **Compatibility:** Males are very intolerant of their own species at all times and become progressively more aggressive towards other males with the approach of spawning. Not given to hyperdominant behaviour when kept in captivity.

Reproductive pattern as outlined for the group. Brood sizes range from 7 to 60 fry. The young are sexually mature at about eight months after release, at 6.5cm(2.5in) for males, slightly less for females. The male's yellow coloration may not be fully developed for another 6 months.

This widely available representative of the *Ps. macrophthalmus* species complex displays considerable variabilty in coloration. Individual males in some local populations are almost entirely yellow. *Pseudotropheus macrophthalmus* 'red cheek' is a small-mouthed species that coexists with *Ps. macrophthalmus* 'yellow head' in the Likoma Island group. As its name implies, the colour in the head and shoulder region of adult males is a deeper orange, while females are a uniform golden yellow. A third, similarly coloured representative of this complex occurs along the southwestern shore of the lake. As its name implies, males of *Ps.*

macrophthalmus 'orange chest' sport an intense yellow-orange flush in front of the belly and on the chest that extends to the opercula and cheeks. Females are greyish brown with seven or eight darker grey vertical bars on the flanks. An extremely attractive orange/black morph of this species is known. These representatives of the *Ps. macrophthalmus* species complex are also commercially available, as are the closely related *Ps. tropheops, Ps. gracilior* and *Ps.* cf. *microphthalmus.*

Pseudotropheus minutus

● **Habitat:** Over the upper surfaces of large boulders and rock slabs. Although its depth range extends to 40m(130ft), it is most abundant in water 3m(10ft) deep or less.
● **Length:** Males up to 10cm(4in); females up to 7.5cm(3in).
● **Sexing:** Young females retain the golden beige base colour of juveniles but, with age, they come to resemble duller versions of the male. Males begin to lose the golden yellow juvenile coloration at 3.8cm(1.5in) and attain their adult coloration at 5.7cm(2.25in).
● **Feeding type:** Herbivore.
● **Compatibility:** As for *P. socolofi* (see page 104).

Above: **Pseudotropheus minutus** ♂ *A territorial male.*

Reproductive pattern as outlined for the group. Females begin breeding between six and eight mont!,is after release, at 5cm(2in). A single spawning can yield from 5 to 50 fry.

This slender mbuna, sometimes known as the 'Likoma Island Elongatus', has largely displaced the true *Ps. elongatus* commercially. That species is one of the few mbuna in which both sexes ferociously defend a territory against allcomers, which doubtless accounts for its rather lurid reputation as an aquarium terrorist! Despite its brilliant coloration, *Ps. elongatus* is not a very satisfactory aquarium resident and its replacement by the equally colourful, but much mellower, *Ps. minutus* is hardly surprising.

Pseudotropheus socolofi

● **Habitat:** An uncommon resident of the rock/sand interface along the eastern coast of the lake at depths of 4-10m(13-33ft).
● **Length:** Males up to 10cm(4in); females slightly smaller.
● **Sexing:** Both sexes share the same powder-blue coloration, but males have larger, more clearly defined and, as a rule, more numerous yellow-orange pseudo-ocelli on their anal fins.
● **Feeding type:** Herbivore.

● **Compatibility:** Males are less aggressive towards other males of their own species than *Ps. lombardoi* and seldom become hyperdominant in captivity. Males defend a territory about 45cm(18in) square, which makes it possible to house more than a single individual in tanks 120-150cm(48-60in) long.

Reproductive pattern as outlined for the group. As many as 75 fry can be produced in a single spawning, but brood sizes of 35-50 are usually the rule. This species does not undergo a sexually related male colour change. The fry are exactly the same colour as their parents and require about eight months to reach sexual maturity, at 6.5cm(2.5in) for males, 5cm(2in) for females.

The adult colour pattern of *Ps. socolofi* is sufficiently similar to that of the undescribed species sold as *Pseudotropheus* 'kingsleyi' to account for confusion over the identity of these two mbuna. Males of *Ps.* 'kingsleyi' rarely exceed 8.3cm(3.25in). They sport an iridescent white or pale yellow spiny dorsal fin and dusky upper and lower distal margins in the caudal. Females are a uniform greyish beige. *Ps. lucerna* is another similarly coloured species, but larger (males up to 15cm/6in) and extremely aggressive.

Above: **Pseudotropheus socolofi** ♂ *Males have larger egg spots.*

Pseudotropheus zebra

● **Habitat:** Areas of broken rock at depths of 5-20m(16.5-66.5ft).
● **Length:** Males up to 12cm(4.75in); females slightly smaller. Captive individuals can reach 15cm(6in).
● **Sexing:** Males have clearly defined egg spots on the anal fin.
● **Feeding type:** Herbivore.
● **Compatibility:** Male *Ps. zebra* defend territories 2m(6.6ft) square in nature. They are very hard on similarly marked mbuna males and apt to become hyperdominant in tanks less than 180cm(72in) long.

Large males are given to post-spawning harassment of egg-carrying females, hence the importance of providing plenty of shelter in their aquarium. Spawns of 30-50 eggs per clutch are normal. Males begin their colour metamorphosis at 4cm(1.6in). It is complete and sexual activity begins about eight months after release, at 7.5cm(3in) for males 6.5cm(2.5in) for females.

Ps. zebra is characterized by both colour polymorphism and the existence of numerous geographic colour variants.

Below: **Pseudotropheus zebra** ♂ *Orange/black aquarium morph.*

Haplochromis and allied genera

The remainder of Lake Malawi's haplochromines are apportioned between 12 genera.
Representatives of seven have been exported over the past 25 years and at least 58 species of *Haplochromis*, 18 species of *Aulonacara*, two species of *Lethrinops* species, two *Hemitilapia* species, as well as the monotypic *Aristochromis christyi* and *Chilotilapia rhoadesi*, have been bred in captivity.
Comprehensive coverage of these cichlids is clearly beyond the scope of this book; the aim has been to present a generally available, representative selection of *Haplochromis* and species of allied genera from several of the main ecological groupings. The great popularity of the 'Malawi peacock cichlids' warrants the inclusion of a wide, though by no means comprehensive, selection of *Aulonacara* phenotypes. All these cichlids require spacious tanks to flourish in captivity.

Aulonacara baenschi

Yellow peacock; sunshine peacock
● **Habitat:** The rock/sand interface in the Chipoka and Maleri Island groups and on the western coast of the lake at Nkhomo. Isolated males and small groups shelter under rock overhangs at depths of 8-20m(27-66ft).
● **Length:** Males up to 9.5cm(3.75in); females up to 7.5cm(3in). Captive males can reach 12.7cm(5in).
● **Sexing:** Males begin to change colour at 3.75cm(1.5in), but it may be another three months before males develop the full intensity of their yellow coloration.

● **Feeding type:** Micropredator. Offer colour foods regularly.
● **Compatibility:** As for *A. stuartgranti* (see page 108).

Sexual maturity is attained six to seven months after release, at 5cm(2in) for males, slightly less for females. Spawns can number from 10 to 40 eggs. Yellow peacock fry are smaller and more delicate than those of *Aulonacara stuartgranti;* careful attention to nitrogen cycle management is essential to rear them successfully.

A longer-snouted *Aulonacara* with a metallic blue head and bright yellow body and fins is sold

Below: **Aulonacara baenschi** ♂ *Colour feeding is essential.*

Above: **Aulonacara freibergi** ♂ *Males are very aggressive.*

under the name yellow-sided peacock. A second, short-snouted species from Usisya, on the northwestern coast of the lake, is known to aquarists as the flavescent peacock. It, too, has a metallic blue head, but its flanks are golden orange rather than bright yellow, and both the dorsal and anal fins sport broad black submarginal bands. Although it may be a colour form of *A. baenschi*, it seems more likely that it represents another undescribed species. Both cichlids are somewhat more aggressive than the yellow peacock under aquarium conditions.

Aulonacara jacobfreibergi
Butterfly peacock
● **Habitat:** The interface between rocky and sandy habitats at numerous localities along the lake's extreme southern coast. Small groups or solitary territorial males shelter under rock overhangs at depths of 4-12m(13-40ft).
● **Length:** Males up to 10cm(4in); females up to 7.5cm(3in). Males grow as large as 15cm(6in) when kept in captivity.

● **Sexing:** Males begin their colour metamorphosis at 2.5cm(1in).
● **Feeding type:** Micropredator. Offer colour food regularly.
● **Compatibility:** As for *A. stuartgranti*, but this species is even more aggressive towards other males of the same species. Large specimens often behave in a hyperdominant manner, even in tanks over 150cm(60in) long.

This species resembles the regal peacock in its sexual precocity. Males can attain full adult coloration and sexual maturity as early as six months after release, at 3.8cm(1.5in). Females begin breeding at the same time, at 3.2cm(1.25in). Spawns of up to 55 eggs have been recorded, but broods of 25-30 fry are more likely.

A closely related, but slenderer peacock species, whose males sport a more deeply notched caudal fin, broader, iridescent white fin margins and well-developed anal fin pseudo-ocelli, has been exported from the southern part of the lake under the names *Aulonacara* 'caroli' and swallowtail peacock. Both these *Aulonacara* species are spectacular show fish and their popularity is fully justified.

Above: **Aulonacara stuartgranti** *Nkhomo Dwarf Regal Peacock.*

Aulonacara stuartgranti
Regal peacock
● **Habitat:** The interface between rocky and sandy habitats at isolated localities along the western coast of the lake. Small groups shelter under rock overhangs at depths of 8-15m(27-50ft).
● **Length:** Males up to 9.5cm(3.75in); females up to 7.5cm(3in). Males can attain 14cm(5.5in) in captivity.
● **Sexing:** Full male coloration appears between six and seven months after release.
● **Feeding type:** Micropredator.
● **Compatibility:** Males are always intolerant of males of their own species and become progressively more aggressive towards other male haplochromines when spawning. They are also likely to become hyperdominant if housed with other *Aulonacara*. Utaka, the specialized zooplankton feeders, and the smaller chisawasawa that live close to sandy substrates, are the preferred companions for this or any other *Aulonacara* species.

Spawns can range from 10-50 eggs, but 20-30 is more normal. Males measuring 3.8-5cm(1.5-2in) at about six months and females of the same age can begin breeding.

Above: **A. stuartgranti** ♀
Female's prominent lateral bars are characteristic of the genus.

While such efforts are usually successful, these early broods rarely number more than 10 fry.

The generic name of these cichlids, literally translated from the Greek, means 'flute-face'. It refers to the enlarged canals of the cephalic lateral line system, which resemble the openings of a flute. The enhanced sensitivity to movement afforded by the expanded lateral line system may allow these cichlids to locate their invertebrate prey more effectively in dim lighting. A second metallic blue *Aulonacara*, characterized by a bright orange-red girdle in the shoulder regions and orange ventral fins, is widely available. Originally sold as the red-shouldered peacock, this species has recently been described and given the name *Aulonacara hansbaenschi*.

Chilotilapia rhoadesi

● **Habitat:** Sandy and intermediate zones. Often associated with *Vallisneria* beds and abundant in water 3m(10ft) deep or less.
● **Length:** Males up to 25cm(10in); females up to 20cm(8in).
● **Sexing:** Males begin to change colour at 11cm(4.3in), 10 months after release. However, full male coloration is not expressed until the fish is two years old.
● **Feeding type:** Micropredator. Live snails are a treat.
● **Compatibility:** As outlined for *H. ahli*, although its requirements for living space are identical to those of *H. venustus* (see page 113). Though less aggressive than males of that species, there are reports of individual male *C. rhoadesi* becoming hyperdominant in a community tank. It is best not to house this fish with species whose females share the combination of an oblique subdorsal stripe and pronounced midlateral band, lest it prove impossible to identify any possible hybrid fry that might result from such cohabitation.

Sexual maturity is attained at about 18 months, at 15cm(6in) for males, slightly less for females. Brood sizes range from 25 to 120 eggs, with clutches of 50 to 60 closer to the norm.

Despite its large adult size, *C. rhoadesi* is such a spectacular show fish that its continued popularity seems assured.

Haplochromis ahli

Electric blue haplochromis
● **Habitat:** A rare deep-water species found over rocks.
● **Length:** Males up to 18cm(7in); females up to 15cm(6in).
● **Sexing:** Males are slow to change colour. The process begins with the appearance of metallic blue in the facial region at 7.5cm(3in) about six months after release and continues for another six months.
● **Feeding type:** Piscivore.
● **Compatibility:** This species is quite capable of keeping fry numbers down in a community tank. Males are violently intolerant of other cichlids with a metallic blue courting dress and quite capable of assuming a

Below: **Chilotilapia rhoadesi** ♂ *A magnificent male.*

Above: **Haplochromis ahli** ♂ *A small but efficient piscivore.*

hyperdominant role if crowded. Does best in tanks more than 150cm(60in) long.

The fish begin spawning at 12 months, but it may take them several months more to attain full reproductive competence. Males court females very forcefully and are inclined to harass them after spawning is complete, so provide plenty of shelter in the breeding tank. Spawns can number 100 eggs, but 50-60 is the average. Fry tend to grow unevenly and are prone to sibling cannibalism. It is important to prevent such behaviour, for *H. ahli* tends to produce broods with a male-biased sex ratio of 3:1 to 4:1. This imbalance is worsened if the smaller females are devoured by their brothers.

This is one of the most spectacularly coloured cichlids to date exported from Lake Malawi. Because males do not attain their full coloration until they are one year old, adult specimens tend to be quite expensive. The nondescript juveniles are much more reasonably priced. Always try to buy a few of the smallest available fry to be certain of obtaining at least one female.

Haplochromis moorii
Malawi blue dolphin
● **Habitat:** This species is one of the chisawasawa, or sandy bottom associated cichlids. Single large individuals or small numbers of juveniles live in close association with inshore foraging groups of sand-sifting cichlids, such as *H. rostratus* or the various *Lethrinops* species.
● **Length:** Males up to 20cm(8in); females up to 16.5cm(6.5in).
● **Sexing:** Unlike most Malawian haplochromines, both sexes of *H. moorii* share the same coloration. However, males usually have a more pronounced hump than females. Their longer soft dorsal and anal fins are a more reliable indicator of sex.
● **Feeding type:** Micropredator.
● **Compatibility:** Despite its adult size, *H. moorii* is a rather unaggressive species that fares poorly when housed with larger or more belligerent tankmates. It does best in tanks more than 150cm(60in) long, in the company of smaller members of its own kind or with the various *Aulonocara* species. With them, *H. moorii* can establish the same close relationship it enjoys in the wild. Not apt to become hyperdominant.

Above: **Haplochromis moorii** ♂♀ *Juvenile male (front) with female.*

Above: ♂ **Haplochromis moorii**
Adult male's nuchal hump.

Males do not always excavate a spawning pit, and courtship is rather low key. Pairs have difficulty coping with more active fishes, so avoid introducing tankmates such as the various mbuna. Parental females do not usually show a prominent throat bulge, but they can be recognized by their conspicuous lateral spots and indistinct vertical barring. Spawns can number 100 eggs, but 60-80 is the average. The blue adult coloration begins to develop at 2.5cm(1in) and is fully developed by six months at 5cm(2in). Reproductive activity begins at about one year, when males are 7.5cm(3in) long.

Haplochromis quadrimaculatus

● **Habitat:** Off rocky shores in areas of local upwelling, known as chirundu, where it finds a plentiful supply of zooplankton. A pelagic species; one of the utaka group of specialized zooplankton feeders.
● **Length:** Males up to 18cm(7in); females slightly smaller.
● **Sexing:** Males begin to change colour at 6.5cm(2.5in), about six months after release. Full male coloration is not attained until 12 months after spawning, and the fish may require an additional two years to reach full adult size.
● **Feeding type:** Micropredator.
● **Compatibility:** As for *H. moorii*, although males are somewhat intolerant of other utaka, particularly if they sport a similar colour pattern. Uncomfortable with mbuna and too small to hold its own with the larger *Haplochromis*

111

Above: **Haplochromis quadrimaculatus** ♂ *A splendid courting male.*

species, *H. quadrimaculatus* does best in the company of small, bottom-living haplochromines, such as the peacocks of the genus *Aulonacara*. One of the few Malawian cichlids to appreciate a fairly deep aquarium.

Males either defend a flat surface as close to the water surface as possible and induce a female to spawn there in the usual haplochromine manner, or they orient their displays to a vertical surface. In the latter case, the pair's spawning behaviour

resembles that of the various *Cyprichromis* species of Lake Tanganyika. Brood sizes range from 20 to 80 eggs, with clutches of 40 to 50 closer to the norm. Sexual maturity is attained at about nine months, at 9cm(3.5in) for males, slightly less for females.

At least a dozen other utaka have been imported, of which half are undescribed. All are superb aquarium residents; their modest adult size recommends them highly to aquarists lacking the tank space to house the larger Malawian haplochromines.

Below: **Haplochromis cf. similis** ♂ *Aptly known as the red empress.*

Above: **Haplochromis venustus** ♂ *An attractive species.*

Haplochromis cf. similis
Red empress
- **Habitat:** Frequently found among *Vallisneria* beds.
- **Length:** Males up to 12.7cm(5in); females slightly smaller.
- **Sexing:** Males begin to change colour at 6.5cm(2.5in), about 6 months after spawning.
- **Feeding type:** Micropredator.
- **Compatibility:** As described for *H. moorii* (see page 110). Although males are somewhat more aggressive than that species, they seldom become hyperdominant in captivity.

Males are sexually mature at about one year. Females begin breeding at about the same time and spawns can number up to 100 fry.

At least three similarly coloured species have been marketed under the name red empress. The fact that none of them can be matched to a described species suggests that the *H. similis* group is in need of competent taxonomic attention!

Haplochromis venustus
Venustus
- **Habitat:** Over inshore sandy substrates to depths of 10m(33ft).
- **Length:** Males up to 20cm(8in); females to 15cm(6in) in nature. Captive individuals are usually somewhat larger.
- **Sexing:** Males grow much more quickly and their colour metamorphosis begins at 10cm(4in), eight months after spawning.
- **Feeding type:** Piscivore.
- **Compatibility:** Like any predator, *H. venustus* will make a meal of any fish it can conveniently swallow. Conversely, it is not a good idea to house this species with other large Malawian piscivores; it is the smallest representative of the group and usually ends up on the losing side of aggressive encounters. Provide a tank at least 150cm(60in) long.

Spawns can number 200 eggs, but the usual range is 80-120. The fry grow rapidly but unevenly, so it may prove necessary to sort them to minimize losses from sibling cannibalism. The young may begin spawning at eight months, but such efforts are rarely successful. Full reproductive competence is typically attained between a year and 14 months after spawning.

This striking species is the smallest of a quintet of closely related benthic predators that consists of *H. polystigma*, *H. linni*, *H. fuscotaeniatus* and *H. livingstonii*. All can be managed in the same manner as *H. venustus*, but need larger tanks.

Index

Page numbers in **bold** indicate major references, including accompanying photographs. Page numbers in *italics* indicate captions to other illustrations. General text and table entries are shown in normal type.

Further reading

Fryer, G. and Iles, T.D. *The Cichlid Fishes of the Great Lakes of Africa* Oliver and Boyd, Edinburgh, 1972

Lewis, D., Reinthal, P., Trendall, J. *A Guide to the Fishes of Lake Malawi National Park* World Wildlife Fund*

Linke, H. and Staeck, W. *Afrikanische Cichliden. I. Buntbarsche aus Westafrika* Tetra-Verlag, Melle, 1981*

Loiselle, P.V. *The Cichlid Aquarium* Tetra Press, Melle, 1987

Ribbink, A.J. et al. A preliminary survey of the cichlid fishes of rocky habitats in Lake Malawi *S. Afr. J. Zool. 18(3): 149 – 310*, 1983*

*Books marked with an asterisk are available from the A.C.A. Book Sales Committee (see below).

Specialist Societies

American Cichlid Association
Membership Committee
P.O. Box 32130
Raleigh, NC 27622
U.S.A.

British Cichlid Association
Membership Secretary
16 Kingsley Road
Bristol BS6 6AF
United Kingdom

A.C.A Books Sales Committee
P.O. Box 423
Cary, NC 27512
U.S.A.

Author's Acknowledgements
The author wishes to thank Ginny Eckstein, Liz and Jim Hutchings, Tom Koziol, Dave Quinn, Delores and Dewey Schehr, Nin Schulz and Dick Strever for their help in the preparation of this book.

Picture credits

Artists
Copyright of the artwork illustrations on the pages following the artists' names is the property of Salamander Books Ltd.

Bill Le Fever: 12-13, 16, 23(part), 24(part), 34, 55, 59
Stephen Gardner: 33
Brian Watson (Linden Artists): 23(part), 24(part)

Photographs
The publishers wish to thank the following photographers who have supplied photographs for this book. The photographs have been credited by page number and position on the page: (B)Bottom, (T)Top, (C)Centre, (BL)Bottom left etc.

David Allison: Endpapers, 39, 51(T), 69, 77, 99
Dr. Chris Andrews: 51
Jan Eric Larrson: Title page, 19, 30-1, 32-3, 44, 46-7(T), 62-3, 64-5, 66, 67, 73, 76(B), 82, 86, 88, 91, 96, 100, 103, 106, 110
Dr. P.V. Loiselle: 17, 18, 71, 72, 75, 79(B), 80, 87, 94, 98, 101, 102, 104, 105, 107, 108, 109, 111(T), 112
Arend van den Nieuwenhuizen: Contents page, 10-11, 57, 60, 68, 76(T), 85, 92-3, 113
Mike Sandford: 79, 90
William Tomey: Half-title page, 35, 50, 83, 84, 89, 111(B)
© Salamander Books Ltd: 20, 26, 28, 30, 46(B), 48

Tilapia buttikoferi